HYDROCARBONS
(ALKANES, ALKENES AND ALKYNES)

HYDROCARBONS

(ALKANES, ALKENES AND ALKYNES)

HYDROCARBONS
(ALKANES, ALKENES AND ALKYNES)

By
Amit Arora

DISCOVERY PUBLISHING HOUSE
NEW DELHI-110002

First Published-2006

ISBN 81-8356-142-X

Published by

DISCOVERY PUBLISHING HOUSE
4831/24, Ansari Road, Prahlad Street,
Darya Ganj, New Delhi-110002 (India)
Phone: 23279245 • Fax: 91-11-23253475
E-mail:dphtemp@indiatimes.com

Printed at:

Amit Enterprises

Preface

This book is written for B.Sc., B.Sc. (Hons.) and M.Sc. students of various universities of India. In this book my aim has been describe the fundamental principles of organic chemistry. Since I do not consider the chemistry of natural products to be fundamental chemistry but rather the application of fundamental principles. The subject matter described in this book covers much of the basic organic chemistry that is needed by a student who wish to study chemistry as a main subject at degree level. The arrangement of the subject-matter is based on homologous series and in general, descriptions of reactions are followed by discussion of their mechanisms and these includes an elementary account of the sort of evidence that led workers to suggest mechanisms that are acceptable at the present time.

We are sure you will appreciate the thought and care that has gone into making of this text. This book has everything you want to have in your advanced organic chemistry text.

The author wish to thank all those who helped in the preparation of the text and illustration, suggestion for improvement and criticism will be gratefully received and acknowledge.

Author

Contents

1

ALKANES

INTRODUCTION

Aliphatic compounds are open-chain or *acylic* compounds, and the name aliphatic arises from the fact that the first compounds of this class to be studied were the fatty acids (Greek : *aliphos* fat)

Carbon forms a large number of compounds with hydrogen only and these are known collectively as hydrocarbons. There are two groups of hydrocarbons.

(i) Saturated hydrocarbons.

(ii) Unsaturated hydrocarbons.

The alkanes or the paraffins are the saturated hydrocarbons. Many occur naturally, and the chief source of the alkanes is mineral oil or petroleum, which occurs in many parts of the worlds.

CLASSIFICATION OF HYDROCARBON

The simplest organic compounds are hydrocarbons containing only carbon-carbon single bonds in their molecules and hydrogen only. In nature hydrocarbons occur chiefly in coal, natural gas and petroleum. All organic compounds can be considered as the derivative of hydrocarbons, obtained by substituting hydrogen both an appropriate functional group.

The compounds of carbon and hydrogen are called hydrocarbons based upon. The nature of bonds reactivity, hydrocarbon can be broadly classified into two main categories (i) Aliphatic (i) Aromatic aliphatic hydrocarbons are be broadly classified into saturated (alkanes) and in saturated hydrocarbon. C Alkenes and alkanes, and acyclic (cycloalkanes) hydrocarbons.

STRUCTURAL FORMULA

The simplest alkane is methane, and its molecular formula is CH_4. Assuming the quadrivalency of carbon and the univalency of hydrogen, we find that there is only one structure possible for methane, viz., (1)

```
     H              Cl             Cl
     |              |              |
  H—C—H        H—C—H        H—C—Cl
     |              |              |
    H(I)          Cl(II)          H(III)
```

Study of the reactions of methane shows that all four hydrogen atoms are equivalent, *e.g.*, methylene dichloride, CH_2Cl_2. prepared by totally different methods is always the same,. Thus (II) and (III) are different ways of writing the same structure. At first sight it may appear that these two structural formula are different. They are different if the molecule is two-dimensional, but saturated compounds the four valencies of car arranged tetrahedrally. Examination of the two structures as tetrahedral figures shows that they are identical. The chief disadvantage of the two-structural formula is that it does not show the spatial arrangement of the atoms. On the other hand, the three dimensional structural formula is cumber some, and for many complicated molecules cannot easily be drawn on paper. Hence we usually adopt the plane-formula when dealing with compounds from the point of view of their structure, only when we which to stress the spatial when we wish to stress the spatial arrangement of the atoms or groups in molecule do we resort to solid diagram. How we show the relative positions of the various atoms or groups in the plane-structural formula of a given compound usually depends on our selves, but where possible the simplest method of writing the structure should be chosen.

Atomic models' are very useful to the organic. Sets of these models may be bought.

Consider 1.5-dichloropentane $C_5H_{10}Cl_2$. Thus is a 'straight' chain compound, and its structural formula is usually written $ClCH_2\ CH_2\ CH_2\ CH_2\ Cl$. Now, 1.5-dichloropentane gives cyclopentane when treated with zinc. Cyclopentane is a ring compound, and to show its formation from 1.5-dichloropentane, we write the structure of the latter as follows.

```
     CH2CH2Cl                    CH2
    /                           /    \CH2
CH2              + Zn ——→ CH2        |     + ZnCl2
    \                           \    /CH2
     CH2CH2Cl                    CH2
```

Thus 'straight' chains may be 'bent' to stress a particular point we may have in mind.

Substitution reactions of methane. Chlorine has no action on methane in the dark. In bright sunlight the reaction is explosive, and hydrogen chloride and carbon are formed.

$$CH_4 + 2Cl_2 \longrightarrow C + 4HCl$$

In diffused sunlight no explosion occurs, but a series of reactions takes place whereby the four hydrogen atoms in methane are successively replaced by chlorine atoms.

$CH_4 + Cl_2 \longrightarrow CH_3Cl + HCl$ methyl chloride

$CH_3Cl + Cl_2 \longrightarrow CH_2Cl_2 + HCl$ methylene dichloride

$CH_2Cl_2 + Cl_2 \longrightarrow CHCl_3 + HCl$ chloroform

$CHCl_3 + Cl_2 \longrightarrow CCl_4 + HCl$ carbon tetrachloride.

In methane the four carbon valencies are satisfied by combination with four hydrogen atoms. Carbon never exhibits a valencies of more than four, and so cannot combine with more than four hydrogen atoms or four other univalent atoms or groups. Hence in the reaction with chlorine, the hydrogen atoms are displaced and chlorine atoms take their place. This type of reaction is known as substitution, and is the direct replacement of hydrogen by some other atom or group The product so formed are known as substitution products. The atom or group that has replaced the hydrogen atom is called the substituent, and when a substituent atom or group is replaced by some other atom or group. the reaction is referred to as replacement (or displacement) reaction. It should be noted that in substitution or replacement reactions there is no change is structure. The spatial arrangement of the molecule, however, may have changed.

STRUCTURAL ISOMERISM IN ALKANES

All the alkanes can be derived formally from methane by substituting hydrogen atoms by methyl groups (CH_3). When the alkane contains three or more carbon atoms, substitution now gives rise to isomerism, *e.g.*, (i) Propane, C_3H_5, can give rise to two butanes. C_4H_{10}, (1) by substitution at a terminal carbon atom, and (1) by substitution at the central carbon atom.

$$\begin{array}{ccccccc} & H & H & H & H \\ & | & | & | & | \\ H- & C- & C- & C- & C- & H \\ & | & | & | & | \\ & H & H & H & H \end{array} \qquad \begin{array}{ccccccc} & & & H & & \\ & & & | & & \\ & H & H- & C- & H & H \\ & | & & | & & | \\ H- & C- & & C- & & C- & H \\ & | & & | & & | \\ & H & & H & & H \end{array}$$

$$\underset{(I)}{CH_3CH_2CH_2CH_3} \qquad \underset{(II)}{CH_3\overset{\overset{\displaystyle CH_3}{|}}{C}HCH_3 \text{ or } (CH_3)_3CH}$$

(I) Has a straight chain, and (II) a branched chain. Both isomers are knows, and thus 'butane' is the first alkane to exhibit structural isomerism. This is an example of chain (or nuclear isomerism, and is characterised by the manner of linking of the carbon chain. (I) is known as normal butane, and (II) as isobutane.

(ii) Three pentanes. C_5H_{12}, may be derived from the butanes.

$$\underset{\text{normal pen tan e}}{CH_3CH_2CH_2CH_2CH_3} \quad \underset{\text{isopen tan e}}{(CH_3)_2CHCH_2CH_2CH_3} \quad \underset{\text{neopen tan e}}{(CH_3)_4C}$$

The bond-diagram way of writing structural formula uses up a lot of a space. Hence it is general practice to use contracted' structural formula, the method is to write the formula in a 'straight' line and enclosing in parentheses any side-chain or substituent group. The atoms or groups in parentheses are joined directly to the preceding carbon atom in the chain not placed in parentheses *e.g.*,

In many cases where there is no ambiguity, the parentheses may be omitted, *e.g.*., $CH_3CH(OH)CH_3$ is often written $CH_3CHOHCH_3$.

As the number of carbon atoms in the alkane increases, the number of possible isomers increases rapidly, *e.g.*, the alkane $C_{15}H_{32}$ can exists in 4347 isometric forms. The number of isomers of a given alkane may be calculated by means of mathematical formula, in most cases very few have actually been prepared.

Examination of the two butane structures shows that all the carbon atoms are not equivalent, and also that all the hydrogen atoms are not equivalent. A primary carbon atom is one that is joined to one other carbon atom, a secondary carbon atom to two other carbon atoms, and a tertiary carbon atom to three other carbon atoms. Hydrogen atoms joined to primary, secondary and tertiary carbon atoms are known as primary, secondary, or tertiary hydrogen atoms, respectively. Thus normal butane contains two primary carbon atoms, two secondary carbon atoms, six primary and four secondary hydrogen atoms. Isobutane contains three primary carbon atoms, one tertiary carbon atom, nine primary hydrogen atoms, and one tertiary hydrogen atom. It will also be seen that normal pentane contains two primary and three secondary carbon atoms, six primary and six secondary hydrogen atoms, and isopentane contains one secondary, one teritary, and three primary carbon atoms, and one tertiary, two secondary, and nine primary hydrogen atoms. Neopentane, however, contains four primary carbon atoms and one quaternary carbon atom, *i.e.*, a carbon atom joined to four other carbon atoms also all the twelve hydrogen atoms are primary. As we shall see later the behavior of these various types of hydrogen atoms differs considerably.

NOMENCLATURE OF THE ALKANES

Whenever a new branch of knowledge is opened up, there is always the problem of introducing a system of nomenclature. The early chemists usually named a compound on the basis of its history, *e.g.*., methane. This is the parent hydrocarbon of methyl alcohol, CH_3OH. Methyl alcohol was originally obtained by the destructive distillation of wood, and was named 'word-spirit'. From this arose the word methyl, which is a combination of two Greek words, methu (wine) and hule (wood). Other examples of this way of naming compounds are acetic acid, which is the chief consituent of vinegar (Latin : *actum,* vinegar), malic acid, which was first isolated from apples (Latin : *malum* apple), and so on. Thus grew up a system of common or trivial names, but in many cases the origin of the name has been forgotten. One advantage of the trivial system is that the names are usually short and easily remembered, but a disadvantage is that a particular compound may have a number of names.

As the number of knows organic compounds increased, it became apparent that it was necessary to systematise the method of nomenclature. This task was originally begun in 1992 by an international committee of chemists a Geneva, and hence is referred to as the Genera system of nomenclature. The work as carried on by the *International Union* of Chemists (I.U.C) by committee appointed in 1922, and in 1931 these drew up a report which is often referred to as the I.U.C, system. Nomenclature is always undergoing revision, and the latest rules are those recommended in 1969 by the Commission on the Nomenclature of Organic Chemistry of the International Union of Pure and Applied Chemistry (I.U.P.A.C). The reader should always consult the *Chemical Society Handbook* is the wished to publish work in the *Journal* , and any mistakes he makes in nomenclature will soon be put right by the Editors!

(1) *In the trivial system :* of nomenclature the straight chain compounds are always designated as normal compounds, and the word normal is usually abbreviated to n. If it contains a quaternary carbon atom, the compound is known as the neo-compound, It is impossible to name many of the more complex alkanes by the trival system (later for examples of the trival system).

The first four alkanes have special names (related to their history), from the fifth member onwards. Latin or Greed numerals are used to indicate the number of carbon atoms in the molecule Table 1.1.

Table 1.1

Name	*Formula*	*Name*	*Formula*
methane	CH_4	hexadecane	$C_{16}H_{34}$
ethane	C_2H_6	heptadecane	$C_{17}H_{36}$
propane	C_3H_8	octadecane	$C_{18}C_{35}$
butane	C_4C_{10}	nonadecane	$C_{19}H_{40}$
pentane	C_5H_{12}	eicosane	$C_{20}H_{42}$
hexane	C_6H_{12}	heneicosane	$C_{21}H_{44}$
heptane	C_7H_{16}	docosane	$C_{22}H_{44}$
octane	C_8H_{18}	tricosane, etc.	$C_{23}H_{48}$, etc.
nonane	C_9H_{20}	triacontane	$C_{30}H_{62}$
decane	$C_{10}H_{22}$	hexatriacontane	$C_{36}H_{74}$
undecane	$CH_{11}H_{24}$	tetracontane	$C_{40}H_{82}$
hendecane		pentacontane	$C_{50}H_{102}$
dodecane	$C_{12}H_{26}$	pentacontane	$C_{60}H_{122}$
tridecane	$C_{13}H_{28}$	heptacontane	$C_{70}H_{142}$
tetradecane	$C_{14}H_{30}$	octacontane	$C_{80}H_{162}$
pentadecane	$C_{15}H_{32}$		

Univalent groups that are formed by the removal of one hydrogen atom from an alkane are knows alkyl or alkyl groups. The name of each individual group is obtained by changing the suffix -one of the parent hydrocarbon. The first four alkyl groups are often represented by a shorthand notation.

The group derived from pentanes was named amyl. This name is now abandoned and pentyl and isopentyl used.

The alkanes, originally and still known as the paraffins, derived this name from the fact that an alkyl group plus one hydrogen atom gives a paraffin. *i.e.*, alkyl H = alkane.

In chemical equations if we are dealing with alkyl compounds as group and we do no wish to specify any particular member, we use the symbol R to represent the unspecified alkyl group, *e.g.*, RCI represents any alkyl chloride.

Table 1.2

Alkane	*Group*	*Short-hand notation*
methane	methyl CH_3 —	Me
ethane	ethyl C_2H_5 --	Et.
propane	n – propyl $CH_3CH_2CH_2$ –	n – Pr, Pr^x, or Pr
	isopropyl CH_3CHCH_3	iso Pr, Pr^β, or Pr^l
butane	n – butyl $CH_3CH_2CH_2CH_2$ –	n – Bu. Bu^x, or Bu
	s – butyl $CH_3CH_2CHCH_3$	s – Bu. Bu^β. Bu^s
	isobutyl $(CH_3)_2$ $CHCH_2$	isoBu or Bu^s
	t – butyl $(CH_3)_3$ C –	t – Bu. Bu^l

(2) In this system of nomenclature the hydrocarbon, except the n-compound, is regarded as a substitution product of methane. The most highly branched carbon atom in the compound is named as the methane nucleus and the alkyl groups attached to this carbon atom are named in alphabetical order. Hydrogen atoms, if joined to the carbon atom chosen as the methane nucleus are not named.

This system of nomenclature is fairly good, since the name indicates the structure of the compound. It is impossible, however to name of the complex alkanes by this system.

(3) In the I.U.P.A.C. system of nomenclature the longest chain possible is chosen and the compound is named as a derivative of this n-hydrocarbon. The carbon chain is numbered from one end to the other by arabic numerals and the positions of side-chains are indicated by numbers, the direction of numbering being so chosen as to give the lowest numbers possible to the side- chains. When series of locants containing the same number of terms are compound term by term, that series is 'lowest' which contains the lowest number on occasion of the first difference, and this principle is used irrespective of the nature of the substituents *e.g.*,

$$\overset{10}{C}H_3\overset{9}{C}H_2\underset{\substack{|\\CH_3}}{\overset{8}{C}H}-\underset{\substack{|\\CH_3}}{\overset{7}{C}H}\overset{6}{C}H_2\overset{5}{C}H_2\overset{4}{C}H_2\overset{3}{C}H_2\underset{\substack{|\\CH_3}}{\overset{2}{C}H}\overset{1}{C}H_3$$

The following are examples of the three system of nomenclature:

Table 1.3

	1	2	3
$CH_3CH_2CH_2CH_3$	n-butane	n-butane	n-butane
1 2 3 4 $CH_3CHCH_2CH_3$ $\quad \vert$ $\quad CH_3$	isopetane	ethyldimethylmethane (C$_2$ is most highly branched)	isopentane
$\quad CH_3$ 1 2| 3 4 $CH_3CCH_2CH_3$ $\quad \vert$ $\quad CH_3$	neohexane	ethyltrimethylmethane (C$_2$ is most highly branched)	2 , 2 dimethylbutane
5 6 7 8 $CH_3CH_2CHCH_2CH_2CH_3$ $\quad \vert$ $\quad ^4CH_2$ $\quad \vert$ $\quad ^3CH_2-CH_3$ $\quad \vert$ $\quad ^2CH-CH_3$ $\quad \vert$ $\quad ^1CH_3$		5-ethyl-2,3-dimethy-octane	

This is named 2.7, 8-trimethyldecane and not 3, 4, 9-trimethyldecane, the first set is 'lower' than the second set because at the first difference 2 it less than 3.

When two sets of numbers are equally possible, then the order of the prefixes in the name decides which shall be used, *e.g.*, I-bromo3-chloropropane and not 3-bromo 1-chloropropane. It should also be noted that the names of prefixes are arranged alphabetically, regardless of the number of each *e.g.*, 5 ethyl 2,3-dimethyl octane.

The I.U.P.A.C. system of nomenclature is undoubtedly superior to the other two, since it permits the naming of any alkane on sight. N.B. The following names are retained for unsubstituded alkanes only, isobutane, isopentane, neopentane, isohexane.

When several chains are of equal length, that chain chosen goes in series to:

(a) The chain which has the greatest number of side-chains,

(b) The chain whose side-chains have the lowest numbered locants,

(c) The chain having the greatest number of carbon atoms in the smaller side-chains,

(d) The chain having the least-branched side-chains. Also where there is a side-chain within a side-chain, the latter is also numbered, and the name of the complex group is considered to begin with the first letter of its complete name *e.g.*,

```
6    5      4      3      2  1
CH3CH——CH——CH——CHCH3
      |      |      |      |
     CH3   CH3   CH3   CH3        4-ethyl-2,3,5-trimethylhexane
             |
            CH3
```

```
                  CH3
                   |1   2     3
              CH3C——CH2CH3
                   |5    4     3      2      1
9    8     7     6
CH3CH2CH2CH2——C——CH2CH2CH2CH3
                   |1   2    3
              CH2CHCH3                 5-(1,1-dimethylprcpyl)-5--(2-methylpropyl)nonane
                      |
                     CH3
```

HOMOLOGOUS SERIES

If we examine the formula of the various alkanes, we find that the formula of each individual differs from that of its 'neighbour' by CH_2, *e.g.*, CH_4, C_2H_6, C_3H_6, C_3H_5, C_4H_{10}, C_5H_{12}, ... A set of compounds, such as the alkanes, in which the members differ in composition from one another by CH_2, is known as an homologous series, the individual members begin known as homologues.

Through organic chemistry we find homologous series, each series being characterised by the presence of a functional group. The functional group is an atom or a group of atoms that causes a compound to behave in a particular way, *i.e.*, it is the functional group that gives rise to homologous series. Some of the more important functional groups and the classes of compounds to which they give rise are shown in Table 1.4.

Table 1.4.

	Functional Group	
Class of Compound	*Formula*	*Name**
Alcohols	—OH	Hydroxyl group
Aldehydes and ketones	$>C=O$	Carboxyl group
Carboxylic acids	$-C(=O)OH$	Carboxyl group
Cyanides	$-N\equiv N$	Cyano group
Nitro-compounds	$-NO_2$	Nitro group
Amines	$-NH_2$	Amino group
Mercaotabs	—SH	Mercapto group
Sulphonic acids	SO_3H	Sulphonic acid group

* Many functional groups are know by more than one name. Nomenclature is delt with in homologous series described in the text.

It is also possible for a compound to contain two (or more) identical or different functional groups and this gives rise to polyfunctional compounds.

If we examine the formula of the various alkanes, we find that the formula C_nH_{2n+2} will represent any particular homologue when n is given the appropriate value, *e.g.*, for pentane n is 5, therefore the formula of pentane is C_5H_{12}. The formula C_nH_{2n+2} is know as the general formula of the alkanes. The composition of any homologous series can be expressed by means of a general formula.

When we study the methods of preparation of the different alkanes, we find that several methods are common to all, *i.e.*, similar methods may be used for the preparation of all the homologues. This gives rise to the general methods of preparation of a particular homologous series.

Examination of the properties of the alkanes shows that many properties are, more or less, common to all the alkane homologues. This rise to the general properties of an homologous series.

The occurrence of homologous series facilitates the study of organic chemistry, since it groups together compounds having many resemblances. If we know the properties of several of the lower homologues we can obtain a fair idea of the properties of higher homologues, *i.e.*, we can forecast (within limits) the properties of a compound that we have not yet prepared. The reason for these limitation is that the behaviour of a functional group in an homologous series depends on structural differences. We can now, however, explain many of these differences in behaviour in terms of polar and steric effects, and we are therefore in a much stronger position to make our predictions.

Only when one functional group is present do the properties of the compounds containing it show many similarities. If two (or more) functional groups are present, the properties depend on the 'distance' between them. When the groups are widely separated, each generally behaves in the normal way, but when they are close to each other, the properties of the compound may be very different from those of the individual functional groups.

There are various ways of studying organic chemistry, *e.g.*, as homologous series, or treating as a class compounds containing a particular functional group, irrespective of the 'rest' of the molecule. In this book, the treatment is the discussion of homologous series.

The General Methods of Preparation of the Alkanes

(1) *Grignard reagents :* Alkyl halides in ether react with magnesium to form alkyl, gnesium halides or Grignard reagents which, on treatment with water or dilute acid, are decomposed to alkanes.

$R - MgX + H_2O$—boil $\longrightarrow$ R—H + Mg CO + DBr for *ex*.

1. $C_2H_5\ Mg\ Br + H_2O \xrightarrow{\text{Boil}} C_2H_6 + Mg\ (OH)$ for ethane.

2. $CH_3\ MgI + H_2O \xrightarrow{\text{Boil}} CH_4 + Mg\ (OH)\ I$ methane.

3. $RI + \xrightarrow{\text{either}} R - Mg - I \xrightarrow{H^+} RH$ (*e.g.*)
4. By heating a mixture of the sodium salt of a carboxylic acid and soda lime.

$$RCO_2Na + NaOH\,(Ca) \longrightarrow RH + Na_2CO_3$$

This process of eliminating carbon dioxide from a carboxylic acid is know as decarboxylate. Soda-lime is a very useful reagent fro this purpose, but cases tested propionate, butyrate and caporate various products were obtained, *e.g.*, with sodium propionate,

$$C_2H_5CO_2Na \xrightarrow{NaOH} \underset{(44\%)}{C_2H_6} + \underset{(20\%)}{CH_4} + \underset{(33\%)}{H_2} + \text{Unsaturated compounds.}$$

This is therefore not useful general method for the preparation of alkanes, since the separation of the products is usually difficult.

When dealing with a preparation, there are at least three important points that must be considered, (i) the yield of crude product, (ii) the yield of pure product, (iii) the accessibility of the starting materials. In certain cases the yield of crude material is high, but the nature of the impurities is such that purification causes a large loss of material, resulting in a poor yield of pure product. On the other hand, it often happens that the product of one reaction is to be used as the starting materials for some other compound which can readily be freed (*i.e.*, purified with very little loss) from the original impurity. Provided, then that this impurity does not interfere with the second reaction, the crude material of the first step can be used as the starting material for the second. Thus the yield alone of a particular reaction cannot decide the usefulness of that method of preparation, the subsequent history of the product must also be taken into consideration. Furthermore, all things being equal, the more accessible materials, *i.e.*, readily prepared or purchased, are used as the starting materials.

(2) By the catalytic reduction of unsaturated hydrocarbons, *e.g.*, reduction of ethylene using nickel. Formerly, the nickel catalyst was usually prepared by reducing with hydrogen nickel

$$R - CH = CH_2 + H_2 \xrightarrow{Ni/573k} RCH_2CH_3 \quad \text{alkane}$$

$$C_2H_4 + H_2 \xrightarrow{Ni} C_2H_6(ex.)^*$$

$$CH_3 - CH_3 = CH_2 \;\; H_2 \xrightarrow{Ni-573k} CH_3 - CH_2 \quad \text{propane}$$

oxide deposited on a suitable inert, porous support, *e.g.*, kieselguhr. The support is impregnated with a nickel salt, treated with sodium hydroxide, washed and dried, and the resulting nickel oxide reduced with hydrogen at 300 – 450°C. Many organic compounds may be reduced by passing their vapours mixed with hydrogen over nickel heated at 200 – 300°C. Any reduction that is carried out in this manner is referred to as the Sabatier-Senderens reduction, in honour of the workers who first introduced this method. It is quite a common feature in organic chemistry to name a reaction after its discoverer or, in certain cases, after a worker who investigated the reaction and extended its application. The reader should always make himself familiar with the reaction associated with a particular name.

A common nickel catalyst is that prepared by the method introduced by Raney (1927). An alloy containing equal amounts of nickel and aluminium is digested with sodium hydroxide, the aluminum, is dissolved away, and the residual very finely divided nickel is washed and stored under water, ethanol, or any other suitable liquid. Raney nickel is more reactive than the supported nickel catalyst, and is usually effective at lower temperatures, often at room temperature

Alkenes may also be reduced by a variety of chemical reagents.

(3) *Reductional of alkyl halides* : This may be done is different ways.

(i) Reduction by dissolving metals, *e.g.*, zinc and acetic or hydrochloric acid, zinc and sodium hydroxide, zinc-copper couple and ethanol , etc. It is no longer considered that 'nascent' hydrogen is the reducing agent. The belief now is that there is an electron-transfer from the metal to the substrate, followed by the addition of protons from the solvent. Thus, reduction with a zinc-copper couple may be formulated as (where e is an electron).

$$Zn \longrightarrow Zn^{2+} + 2e$$

$$RX + e \longrightarrow X^- + R. \longrightarrow R^{:-}$$

$$R:^- + C_2H_5OH \longrightarrow R-H + {}^-OC_2H_5 \text{ (g, vg.)}$$

(ii) Primary and secondary alkyl halides are readily reduced to alkanes by lithium aluminium, hydride ($LiAIH_4$), tertiary halides give mainly alkenes. On the other hand, sodium borohydride ($NaBH_4$) reduces secondary and tertiary halides, but not primary, whereas triphenyltin hydride (Ph_3SnH) reduces all three types of halides.

(iii) Catalytic hydrogenolysis

(iv) Iodides may be reduced by heating with concentrated hydriodic acid at 150°C.

Reduction with concentrated hydriodic acid is usually carried out in the presence of a small among of red phosphorus which regenerates the hydroids acid from the iodine formed. The hydriodic acid red phosphorus mixture is one of the most powerful reducing agents used in organic chemistry.

Alcholos, aldehydes, ketones, and carboxylic acids are also reduced to alkanes by hydrogen iodide and red phosphorus. On the other hand, aldehydes and ketones may be reduced to alkane by the Clammensen.

(4) *Wurtz reaction (1854) :* An ethereal solution of an alkyl halide (preferably the bromide or iodide) is treated with sodium, *e.g.*,

$$R^1X + R^2X + 2Na \longrightarrow R^1 - R^2 + 2NaX$$

As previously pointed out, when we do not wish to specify a particular alkyl group, we use the symbol R. When we deal with two unspecified alkyl groups which may, or may not, be the same, we can indicate this by R^1 and R^2, also when dealing with compounds containing a halogen atom, and we do not wish to specify the halogen, we can indicate the presence of the unspecified halogen atom by means of X.

Consideration of the equation given above shows that in addition to the desired alkane $R^1 - R^2$, there will also be present the alkanes $R^1 - R^1$ and $R^2 - R^2$. Unsaturated hydrocarbons are also obtained. Obviously, then the best yield of an alkane will be obtained when R^1 and R^2 are the same, *i.e.,* when the alkane contains an even number of carbon atoms and is symmetrical. It has been found that the Wurtz reaction gives good yields only for 'even carbon' alkanes of high molecular weight, and that the reaction generally fails with tertiary alkyl halides (q, v).

Sodium is used in the Wurtz reaction. Other metals, however, in a finely divided state, may also be used, *e.g.*, Ag Cu.

Tow mechanisms have been suggested for the Wurtz reaction.

(i) The intermediate formation of an organo-metallic compound, *e.g.*, the formation of n-butane from ethyl bromide.

$$C_2H_5{-}Br + 2Na \longrightarrow C_2H_5{-}Na^+ + NaBr$$

$$C_2H_5^- Na^+ + C_2H_5Br \longrightarrow C_2H_5 - C_2H_5 + NaBr$$

(ii) The intermediate formation of free radicals, *e.g.*,

$$C_2H_5 - Br + Na \longrightarrow C_2H_5 + NaBr$$

$$C_2H_5 + C_2H_5 \longrightarrow C_2H_5 - C_2H_5$$

On of the properties of free radicals is disproportionation, *i.e.*, intermolecular hydrogenation, one molecule acquiring hydrogen at the expense of the other, *e.g.*,

$$C_2H_5 + C_2H_5 \longrightarrow C_2H_6 + C_2H_4$$

This would account for the presence of ethane and ethylene in the products. According to Morton *et al.* (1942). However, ethane and ethylene may be produced as follows:

$$Na^*CH_3CH_2 H{-}CH_2{-}CH_2{-}Br \longrightarrow CH_3CH_3 + CH_3{=}H_2 + NaBr$$

The is a bimolecular elimination mechanism. On the other hand, the mechanism of alkane formation is generally accepted as usually being a bimolecular nucleophilic substitution.

$$Na^*CH_3CH_2 H{-}CH_2{-}CH_2{-}Br \longrightarrow CH_3CH_3 + CH_3{=}H_2 + NaBr$$

The free- radical mechanism, however, has been shown to operate in some cases (Bryce Smith, 1963).

(5) *Kolbe's electrolytic method (1849) :* A concentrated solution of the sodium or potassium salt of a carboxylic acid or mixture of carboxylic acids is electrolysed. *e.g.*,

$$R^1CO_2K + R^2CO_2K + 2H_2O \longrightarrow R^1{-}R^2{+}2CO_2{+}H_2{+}2KOH$$

$$R - COO\ Na\ (ag) \xrightarrow{\text{dissiociation}} Na^+ + R - COO^-$$

At Anode

$$R\ COO^- \longrightarrow R\ COO^+ + C$$

$$R\ COO^- \longrightarrow R^+ + CO_2$$

$$2R^+ \longrightarrow R - R \text{ higher altitude}$$

At Cathode

The Na– ions carry the current Through the solution but do not get reduced at cathode Instead H^+ ions produced due to one, ionsection of water are invalued in two electrode reaction at cathode. The reaction taking place at cathode are.

$$2H_2O \rightleftharpoons 2H^+ + 2OH^-$$

$$2H^+ + 2e^- \longrightarrow H_2\,(g)$$

Not reaction at cathode. $2H_2O + 2e^- \longrightarrow H_2\,(g) + 2OH^-$

If R^1 and R^2 are different, then hydrocarbons $R^1 - R^1$ and $R^2 - R^2$ are also obtained (of. Wurtz reaction). Such mixtures can often be separated readily. Yields of 50–90 per cent have been obtained with straight chain acids containing 2–18 carbon atoms. Alkyl groups in the a-position decrease the yield (usually below 10 percent). The by products are alkenes, alcohols (particularly in alkaline solution) and esters. It is also interesting to note that the yields of the alkanes are increased when dimethyl formamide is used as solvent (Finkelstien *et al.*, 1960).

The Kolbe's electrolytic method now has application in the synthesis of natural compounds, particularly lipids.

Several mechanisms have been proposed for the Kolbe's reaction. The free radical theory is the on now favoured, and strong support for it has been provided by Lippincott *et al.* (1956), *e.g.*, where sodium propionate is electroysed, n butane, ethane, ethylene and ethyl propionate are obtained. The propionate ion discharges at the anode to form a free radical.

$$C_2H_5CO_2 \longrightarrow C_2H_5CO_2 + e$$

This propionate free radical then breaks up into the ethyl free radical and carbon dioxide

$$C_2H_5CO_2 \longrightarrow C_2H_5 + CO_2$$

Then (i) $2C_2H_5 \longrightarrow C_4H_{10}$

(ii) $C_2H_5 + C_2H_5 \longrightarrow C_2H_5 + C_2H_4$

(iii) $C_2H_5 + C_2H_5CO_2 \longrightarrow C_2H_5CO_2C_2H_5$

Reaction (i) given n-butane, (ii) gives ethane and ethylene by disproportionation (ef. Wurtz reaction), and (iii) gives ethyl propionate.

(5) A very useful method for preparing long-chain alkanes involves the coupling of alkylboranes by means of silver nitrate in the presence of sodium hydroxide at 25°C, *e.g.*, 2-methylpentene gives 4,7-dimethyldecane (Brown et at., 1961).

$$6CH_3(CH_2)_2C(CH_3)=CH_2 \xrightarrow{2B_2H_4} [2CH_3(CH_2)_2$$

$$B \xrightarrow[NaOH]{AGNO_5} 3CH_3(CH_2)_2CH(CH_3)CH_2CH_2$$

$$CH(CH_3)(CH_2)_2CH_3 \qquad (75\text{-}80\%)$$

Coupling may also be effected between unlike boranes.

Conformation. It has already been pointed out that the valencey angle at the central carbon atom of a regular tetrahedron is 109°28'. It was originally postulated (van't Hoff). Another postulate was fixed but it is now known that it may deviate from this value (± 5°). Another postulate was the principle of free rotation about a single bond. Let us consider the ethane molecule, $CH_3 - CH_3$, and let us imagine that one methyl group is rotated about the C – C bond as axis with the other group at rest. Suppose we use, as the starting point, the position in which two C – H bonds are parallel, *i.e.*, these four atoms lie in a plane (Fig. 1.1c).

In this position, the *dihedral angle* (*angle of rotation or angle of torsion*) is zero, and if the rotation is free, then as the dihedral angle changes, the energy content of the molecule will remain constant (the plot of energy content against the dihedral angle will be a horizontal line). In this situation, the two haves can assume with complete freedom, and infinite number of positions relative to each other. Thus the entropy of the molecule will be a maximum.

Pitzer *et al.* (1936) calculated the entropy of ethane based on the assumption that the internal rotation was free, and found that calculated value was greater then the observed value. This means that there is less freedom to assume all possible dihedral angles than was expected on the principle of free rotation. Pitzer therefore assumed that the internal rotation is hindered by a potential energy barrier, and calculated the change in entropy with increasing barrier height.

When he assumed a barrier of 12.55kj mol^{-1} (3k cal), the calculated and observed entropies were brought into good agreement. The potential energy curve obtained for ethane is shown in (Fig. 1.1a) and by convention the potential energy is measured relative to the energy of the most stable form.

Because of the existence of potential energy barriers, the result is *mutual oscillation* (*libration*) about the conformations with minimum potential energy, there by producing 'more order' (more restriction) than had there been no barriers.

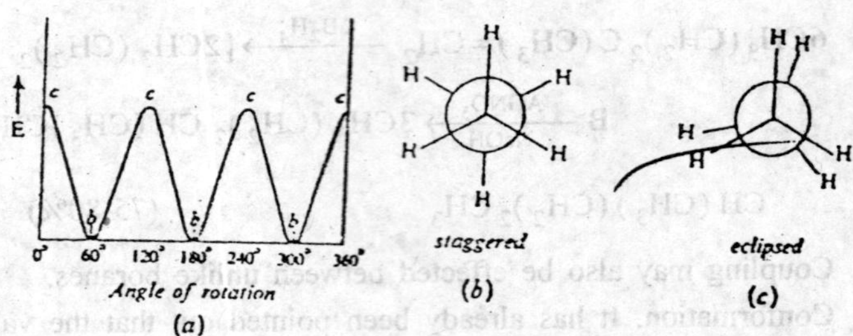

Fig. 1.1

Fig. 1.1b is the Newman projection formula. This is obtained by viewing the molecule along the bonding line of the two carbon atoms, with the carbon atom nearer to the eye being designated by equally spaced radii, and the carbon atom further from the eye by a circle with three equally spaced radial extensions.

The *staggered* (or transoid) conformation (in which the hydrogen atoms are as far apart as possible), and Ic the *eclipsed* (or cisoid) conformation (in which the hydrogen atoms are as close together as possible). It can be seen from that the energy of the eclipsed conformation is greated than that of the staggered conformation.

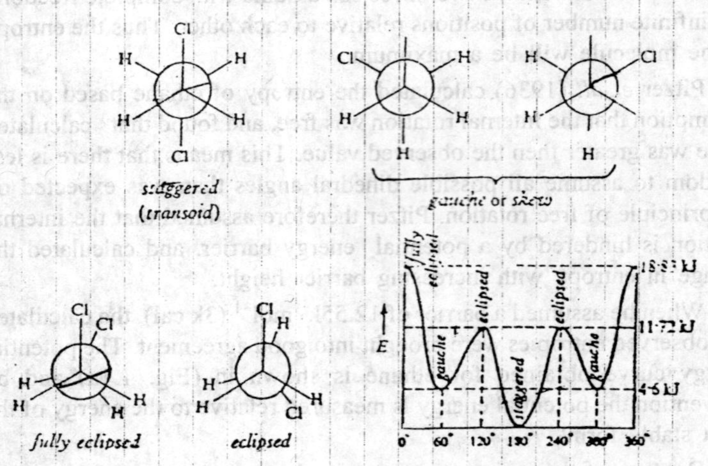

Fig. 1.2

The potential energy barrier (11.92 kj mol^{-1}) is much too small for either conformation to remain stable, *i.e.*, the eclipsed and staggered forms are readily inter convertible and hence neither can be isolated as such. However the staggered conformation is the preferred from, *i.e.*, its population is greated than that of the eclipsed form.

Now let us consider ethylene dichloride. According to Bernstein (1949), the potential energy undergoes the changes shown in Fig. 1.2. There are two positions of minimum energy, but the staggered conformation is the more preferred one, *i.e.*, the one in which the molecule largely remains.

Table 1.5

Z	CH_3Z τ	$R'CH_2Z$ τ	R_2CHZ τ	Z	CH_3Z τ	$R'CH_2Z$ τ	R'_2CH_2 τ
—R	9.12	8.75	8.50	—I	7.83	6.85	5.78
$—CO_2R$	8.00	7.90	—	—Br	7.35	6.70	5.97
—CN	7.98	7.52	7.3	—Cl	6.98	6.56	5.98
$—CONH_2$	7.98	7.95	—	—OR	6.70	6.64	6.20
$—CO_2H$	7.93	7.65	7.43	$—NR_3$	6.67	6.60	6.50
—COR	7.90	7.60	7.5	—OH	6.62	6.42	6.15
—SH, —SR	7.90	7.60	6.90	—OCOR	6.35	5.89	4.95
$—NH_2$,$—NR_2$	7.85	7.50	7.13	—F	5.70	5.66	5.40
—CHO	7.83	7.80	7.60	$—NO_2$	5.67	5.60	5.40

	τ
R—OH	9.5 – 6.0 (lower for enols; – 1 to –6)
Ar—OH	~5.5
R—SH	9–8
Ar—SH	~6.5
RNH_2	8.5
$ArNH_2$	6.6 – 5.0
RCO_2H	0 to –3

J values (Hz)

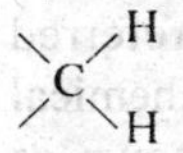

$>C<^{H}_{H}$ 10–18 $>CH—CH<$ 0–12 (no rotation)

dihedral angle. 0°, ~ 8; 90°, ~ 0; 180°, 9 – 11

Dipole moment studies show that this is so in practice, and according to Mizushima *et al* (1038), only the staggered from is present at low temperatures.

Thus, in theory, there is no free rotation about a single bond, but in practice it may occur the various forms do not differ very much in energy content (usually between 4.18 and 41.84kj mol^{-1}). Molecules which can form isomers by rotation about single bonds are called flexible molecules and the different forms taken up are different conformations or rotational isomers. Usually, the staggered conformation is the most stable eclipsed form is always avoided where possible.

Free rotation about a single bond is generally accepted in *simple* molecules. *Restricted rotation* about a single bond, however, may take place when the molecule contains groups large enough to impede free rotation, *e.g.*, in *ortho*-substituted biphenyls. In some cases resonance may give rise to restricted rotation about a single bond.

Energy barrier heights have been determined by various methods, *e.g.*, infra-red, Raman and NMR spectra, dipole moment measurements, etc. The origin of these potential barriers is complex, several types of contributing intersections operate simultaneously, their relative importance depending on the structure of the molecule in question. Four factors are:

(i) Dipole-dipole interactions, poles of like sign will tend to be as far apart as possible.

(ii) Bond opposition strain, Covalent bonds, because they 'consist' of electron pairs, will repel other bonds

(iii) Steric strain, *i.e.*, repulsion between non-bounded atoms or groups. When approach closely together, forces of repulsion come into play, and it has been found that the distance of closest approach is greater than the sum of the covalent radii of the atoms concerned. This distance of closest approach for each atom (or group) is known as the can der *Walls radius*, *e.g.*, H, 1.2Å, O 1.4Å, Cl, 1.8Å, Br, 1.95Å, I 2.13Å.

(iv) Bond angle strain, *i.e.*, distortion of bond angles.

When the stability of a molecule is decreased by internal forces product by interaction between constituent parts, that the molecule is said to be under strain. The study of the existence of preferred conformations in molecules, and the relating of physical and chemical properties of a molecule to its preferred conformation, is known as conformation analysis.

A particularly useful method for studying the conformations of molecules is NMR spectroscopy. However, before dealing with this, let us first consider the chemical shifts of the protons of CH_3, CH_2, and the CH group in acyclic compounds and the chemical shift of a proton attached to O, N, etc.

As we have seen (p, 11), the more electronegative Z is in the groups – CH – Z and – Z – H, the more is the proton desheilded and consequently the lower is the value lists the values of some of the more important groupings, and also lists some coupling constants.

We return now to the use NMR spectroscopy in conformational analysis. Let us first consider ethyl chloride, and since the most stable conformation is the staggered one, the molecule may be represented as (I), (II), and (III), In (I), the environments of H_a and H_c are identical,

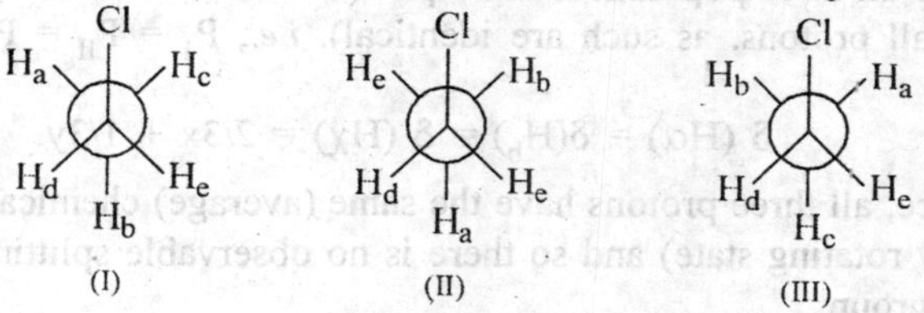

Fig. 1.3

but differ from that of H_b. This can be seen to be the case by replacing each proton, one at a time, by Z. This procedure for H_a and H_c produces mirror images, and so the two protons are chemically equivalent, but for H_b the substituted product is different because of the different dihedral angles, and similarly for H_b. Thus, H_a and H_c are chemically but not magnetically equivalent.

By using similar arguments, it can be shown that H_a and H_b are also chemically but not magnetically equivalent. Hence, if the population of ethyl chloride conformation were completely represented by (1), protons Ha and H_b would give one signal and H_a another signal, provided that the chemical shifts were sufficiently different (0.1 – 0.2 p. pm.), *i.e.*, the methyl group gives would two signals. In practice, the methyl group gives only one signal a (a triplet), and therefore protons H_a, H_b, and H_b must be equivalent.

This is explained on the basis that there is free rotation about the carbon-carbon single bond. When the methyl group rotates (with respect to the CH_2CI group), it can take up the other two stable staggered conformations, (II) and (III), the result being that each proton has an average environment. Since these average environments are identical, rotation result in the equivalence of H_a, H_b, and H_c.

This equivalence may be demonstrated as follows. Let P_1, P_{II}, P_{III} be the population in conformations (I), (II), and (III), respectively, where $P_1 + P_{II} + P_{III} = 1$. Also let δ (the chemical shift) of any proton trans (anti) to another proton be x, and y if *trans* to Cl. Now, the observed δ of a given proton is the sum of the δ-contributions in each conformation, and so it follows that,

$$\delta\ (H_a) = P_1x + P_{II}y + P_{III}x$$

$$\delta\ (H_b) = P_1y + P_{II}x + P_{III}x$$

$$\delta\ (H_c) = P_1x + P_{II}x + P_{III}y$$

Since all three populations are equal (all three are indistinguishable because all protons, as such are identical), *i.e.*, $P_1 = P_{II} = P_{III} = 1/3$, therefore

$$\delta\ (H\alpha) = \delta(H_b) = \delta\ (H\chi) = 2/3x + 1/3y.$$

Hence, all three protons have the same (average) chemical shift (in the freely rotating state) and so there is no observable splitting for the methyl group.

Application of this method to protons H_d and H_e gives.

$$\delta\ (H_d) = P_1x + P_{II}x + P_{III}x = x$$

$$\delta\ (H_e) = P_1x + P_{II}x + P_{II}Ix = x$$

From the above discussion, it can be seen that if ethyl chloride were undergoing slow rotation, it would be an ABB.CC spin system, but when undergoing fast rotation, it rotation, it is an A_3B_2 spin system (see also below).

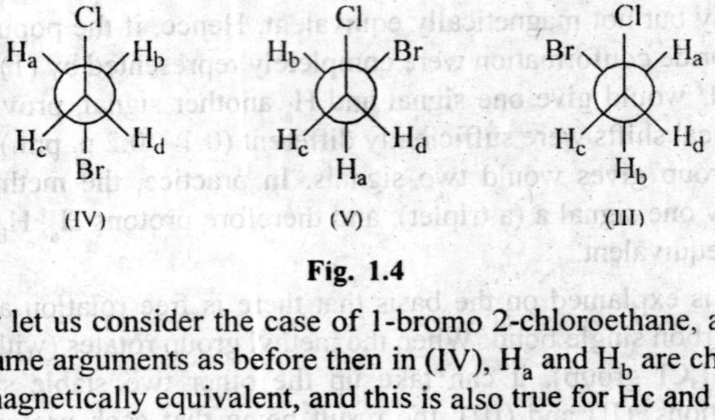

(IV) (V) (III)

Fig. 1.4

Now let us consider the case of 1-bromo 2-chloroethane, and if we use the same arguments as before then in (IV), H_a and H_b are chemically but not magnetically equivalent, and this is also true for Hc and H_d. Also if δ for a proton trans to a proton is x, trans of CI is y, and trans to Br is z, then,

$$\delta\,(H_a) = P_{iv}x + P_v y + P_{vi}x$$
$$\delta\,(H_b) = P_{iv}x + P_v x + P_{vi}y$$

In this molecular, although $P_v = P_{vi} \neq P_{iv}$, it still follows that $\delta\,(h_a) = \delta\,(H_b)$. In the same way, it can be shown that $\delta\,(H_c) = \delta\,(H_d)$.

We can extend the argument by also by also considering the possible coupling constants. If J_t and J_t represent, respectively, the vicinal trans and skew coupling, then,

$$J_{ad} = P_{iv}J_t + P_v J_x + P_{vi}J_x$$
$$J_{bd} = P_{iv}J_x + P_v J_t + P_{vi}J_x$$

Since $P_v = P_{vi} \neq P_{iv}$ therefore $J_{ad} \neq J_{bd}$. Hence H_a and H_b (and Hc and H_d) are chemically but not magnetically equivalent. Application of this method will show that the three protons in the methyl group in ethyl chloride are magnetically equivalent, as are also the two protons in the CH2CI group, for the freely rotation molecule.

Finally, let us consider a molecule of the type $R^1R^2CHCH^3R^4$. Each carbon atom is asymmetric and the environments of H_a are different in all three conformations, and this is also true for H_d.

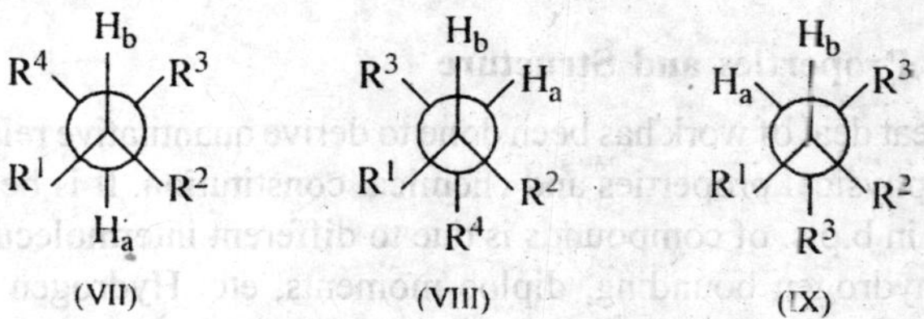

Fig. 1.5

Furthermore, because of the different steric effects, the three populations will be different (*i.e.* $P_{vii} \neq P_{viii} \neq P_{ix}$). Hence, the chemical shifts of H_a (and those of H_b) are different in each conformation. If rotation were slow, the NMR spectrum would be a composite of all three spectra (three AB spin systems for each pair of enantiomers). If the temperature is raised, the rate of rotation is increased, and the result is an average of chemical shifts and coupling constants to given one AB spin system (for each pair of enantiomers). These averages, however, will be weighted in favour of the most stable conformation, *i.e.*, the one with the highest population.

The name paraffin arose through contracting the two Latin words parum affinities, which means little affinity. The name was suggested because these hydrocarbons were apparently very reactive. It is difficult to define the terms 'reactive' and unreactive, since a compound may be reactive under one set of conditions and unreactive under another. Under 'ordinary' conditions, the alkanes are inert towards reagents such as acids, alkalis, oxidising reagents, reducing reagents, etc, but are reactive if the 'right' conditions are uses (see below). One possible reason for this lack of reactivity is that the C – H bond is very strong.

PHYSICAL ALKANES

The normal alkanes from C_1 to C_4 are colorless gases, C_3 to C_{17}, colorless liquids, and from C_{18} onwards, colorless solids. The b.p.s. rise fairly regularly as the number of carbon atoms in the compound increases. This holds good only for the normal compounds, and the difference in b.p.s. decreases as the higher homologues are reached. Other physical properties, such as m.p., density, viscosity, also increases in the same way as the b.p.s. (of the normal alkanes), *e.g.*, the density of the normal alkane increases fairly steadily for the lower members, and eventually tends to a maximum value of about 0.79.

Physical Properties and Structure

A great deal of work has been done to derive quantitative relationships between physical properties and chemical constitution. It is believed the variation in b.p.s. of compounds is due to different intermolecular forces such as hydrogen bounding, diploe moments, etc. Hydrogen bounding may produce association, and this will cause the b.p., to be higher than anticipated. The greater the dipole moment of the compound, the higher is the b.p., since owing to the charges, more work is required to separate then molecules, *e.g.*, nitro-compounds, RNO_2, which have large dipole moments, have much higher b.p.s. Than the alkanes in which the dipole moment is absent or very small. As far as the alkanes are concerned, the only attractive forces are the weak van der Waals forces.

Observation has shown that in group of isomeric compounds (acyclic), the normal compound always has the highest b.p., and m.p., and generally, the greater the branching the lower is the b.p. In the n-compound, owing to its large surface area due to the linear nature of the chain, the van der Waals attractive forces are relatively large. As branching increases, the molecule tends to become spherical, thereby resulting in a decreased surface area available for intermolecular attractions. Furthermore, because intermolecular forces are small in alkanes, the b.p.s. of these compounds are used as reference values for given molecular weights, *e.g.*, propane (M.W. 44, b.p. – 44.5°C), ethanol (M.W. 46, b.p. 78.1°C). The high b.p., of the latter is attributed to hydrogen bounding.

Solubility

The alkanes are almost insoluble in water, but readily soluble in ethanol and ether, the solubility diminishing with increase in molecular weight.

It is believed that solubility depends on the following intermolecular forces, solvent/solute, solute/solute, solvent/solvent. If the solvent/solute attractive forces are greater than either of the other two, solution will be expected to occur fairly readily. A non-electrolyte dissolves readily in water only if it can form hydrogen bonds with the water. Thus alkanes are insoluble or almost insoluble, in water. Methane is more soluble than any of its homologues, hydrogen bonding with the water is unlikely, and so other factors possibly molecular size–must also play a part. Useful rule in organic chemistry with respect to solubility is that like dissolves like. *e.g.*, if a compound contains a hydroxyl groups, then the best solvents usually contain hydroxyl groups.

X-ray analysis of solid alkanes has shown that the carbon chains are fully extended, i.e, zigzag. In the liquid state this extended form is also one of the stable conformations provided that the carbon chain is not very long. When the number of carbon atoms is sixteen or more, the extended form is no longer present in the liquid state.

Since the dipole moment of methane is zero, the dipole moment of the methyl group is equal to that of the fourth C – H bond (in methane) and is directed along this axis. Thus replacement of hydrogen by a methyl group will not be expected to change the dipole moment, *i.e.*, the dipole moment of all alkanes, whether straight -or branched-chain, will be zero. Thus has been found to be so in practice. This will always hold good whatever conformation is taken up by the alkane provided that no deformation of the normal carbon valencey angle (of 109°.28') is produced in the twisting, since all methyl groups will be balanced by a C – H bond. It therefore follows that the electron negativity of all alkyl groups is equal to that of hydrogen, namely zero. As soon as one hydrogen atom is replaced by another atom of group (other than alkyl), the resultant molecule will now be found to possess a dipole moment.

The position of the infra-red absorption region of the CH group (stretch) in alkanes depends on whether the carbon atom is primary, secondary or tertiary. Thus we have the following regions: —CH_3, 2975 – 2950 and 2885 – 2860 cm^{-1}(m); —CH_2—, 2940 – 2915 and 2870 – 2845 cm^{-1}(m); >CH, 2900 – 2880 cm^{-1}(w). SOme C—H deformation absorption frequencies are: CH_3, 1470–1435 (m) and 1385 – 1370) cm^{-1} (s), – CH_2 – 1480 – 1440 cm^{-1} (m).

Two useful skeletal vibrations are is, Me_2CH – 1175–1165cm^{-1} (s), neo, Me_3C – 1255 – 1245 cm^{-1} (s). It is thus possible to detect the presence of these groups in a molecule.

The infra-red spectrum of 3-methylpentane (as a film). These bands are characteristic of all types of alkanes containing methyl and methylene groups.

Mass Spectra

Since the stability of carbonium ions is in the order t > s > prim, fission of bonds in alkanes occurs preferentially at branched carbon atoms. When alternative fissions can occur, it is the heaviest side-chain that is eliminated preferentially. Since alkyl carbonium ions are formed, all of them (with ^1H and ^{12}C) will give peaks of odd masses. In particular, n-alkanes give a series of peaks separated by 14 mass units (CH_2). The relative abundance of these peaks is usually greatest for $C_3H_7^+$ (43), C (57), and $C_5H_7^+$ (71), and decreases fairly regularly for the larger masses. Furthermore, each peak is generally accompanied by peaks of mass 1 and 2 units lower, corresponding to the loss of 1 and 2 hydrogen atoms, respectively.

The molecular ion is always present for n-alkanes, its intensity decreasing with increasing molecular weight. On the other hand, the greater the branching in the alkane, the less is the likelihood of the appearance of the molecular ion and, if this does appear, its intensity is usually low.

PHYSICAL PROPERTIES OF ALKANES

Some of the physical properties that we will study in this section are boiling point, melting point density and solubility. All these properties depend upon the intermolecular forces between the molecules. These forces are of the van der Walls' type. The van der Walls' forces depend upon the shape and structure of the molecules.

A straight chain molecule has more are of contact with an adjacent molecule as compared to a branched chain molecule which is more compact. Therefore, *the van der Waals' forces are greater in straight chain molecules than in branched chain molecules*.

Major physical properties of alkanes are described below.

1. *Physical state :* Lower alkanes containing upto four carbon atoms (C_1 to C_4) are colorless, odourless gasses. The alkanes with five to sixteen carbon atoms ($C_5 - C_{16}$) are colourless and odourless liquids, and the higher members ($> C_{18}$) are colourless (white) waxy solids at near room temperature.
2. *Boiling point :* Alkanes are non-polar molecules. The molecules of alkanes are bond to each other by weak van der Waals' forces, and therefore, have lower boiling points.

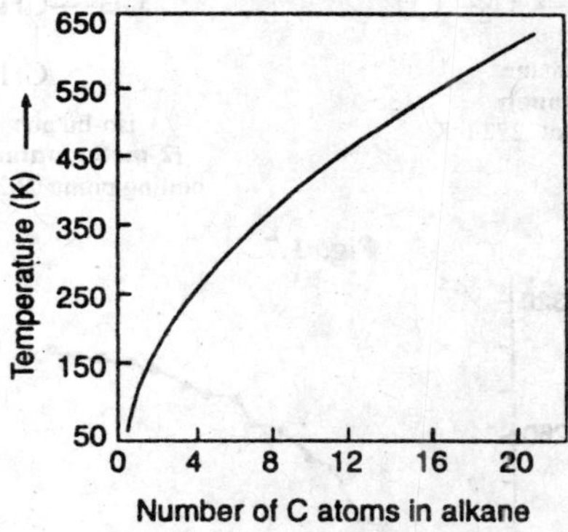

Fig. 1.6 : The Variation of the boilling points of n-alkanes.

(a) The boiling points of alkanes show a regular increases with the number of carbon atoms in the alkane. The boiling point increases by about 20–30 degrees for each addition of a – CH_2 group to the molecular formula. Variation of the boiling points of some n-alkanes is shown in Fig. 1.6.

Explanation. With an increases in the molecular size/mass, the strength of the intermolecular forces (van der Waals' forces) increases. The boiling point of a compound depends upon the strength of the intermolecular forces between the molecular. Therefore, the boiling points of alkanes increase with an increase in the chain length.

(b) *The branched chain isomers of an alkane have lower boiling points than the straight chain isomer.*

Explanation : This is because the branched-chain isomers are more compact and tend to acquire nearly spherical shape. This decreases the effective surface are of contact with the adjacent molecule. As a result, the strength of the van der Waals' forces in the branched chain alkanes are weaker than that in the straight chain alkanes. For example,

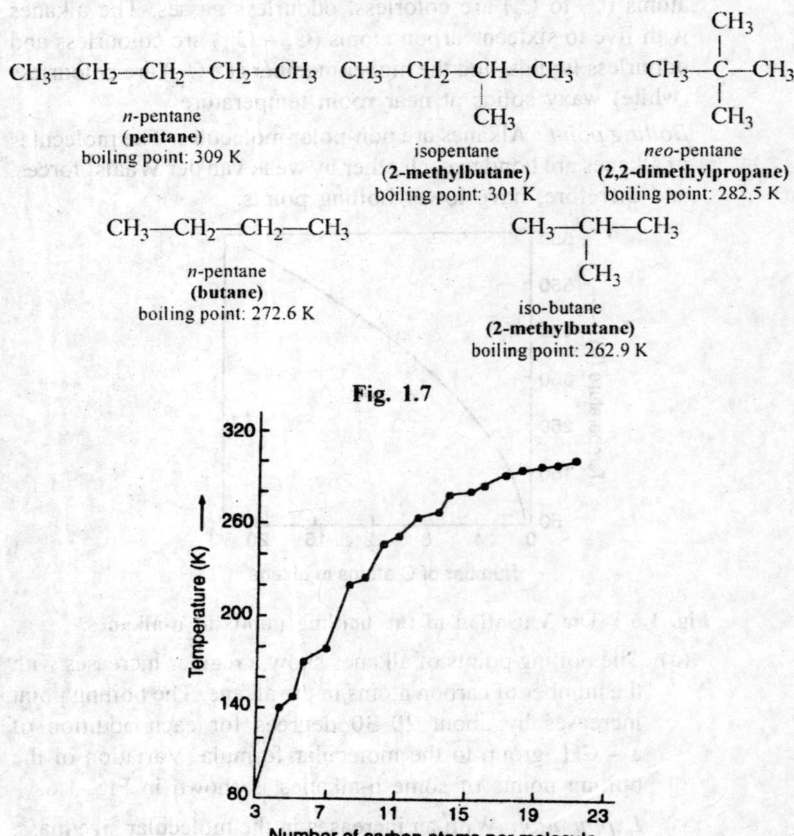

Fig. 1.7

Fig. 1.8 : Variation in the melting points of n-alkanes with increase in the number of carbon atoms.

3. *Melting point :* The melting points of n-alkanes do not show a regular variation. For lower members, the alkanes with even number of carbon atoms have higher melting point than the next higher or lower alkane having odd number of carbon atoms.

   However, four alkanes with four or more carbon atoms, the increase in the melting point in going from odd to even alkane, is mush higher than that observed in going from even to odd alkane. The variation of melting point with number of carbon atoms in the chain of some n-alkanes is shown in Fig. 1.8.

The melting point of some alkanes are given in Table 1.6.

Table 1.6 : The melting points os some alkanes.

Alkane		CH_4	C_2H_6	C_3H_8	C_4H_{10}	C_5H_{12}	C_6H_{14}	C_7H_{16}	C_8H_{18}	C_9H_{20}	$C_{10}H_{22}$
Melting point,	°C	− 186	− 172	− 190	− 135	− 130.5	− 94.3	− 90	− 56.5	− 51	− 32
	K	87.1	101.1	83.1	138.1	142.6	178.8	183.1	216.6	222.1	241.1

The observed variation of melting point with the number of carbon atoms can be explained as follows.

Explanation : The carbon chains in alkanes are zig-zig because the carbon atoms do not lie in the same vertical plane Fig. 1.9.

Fig. 1.9 : The zig-zag chain of carbon atoms in alkanes.

For an alkane consisting of even number of carbon atoms the terminal methyl groups lie on the opposite side. For alkanes consisting of odd number of carbon atoms, the terminal methyl groups (hence carbon atoms) line on the same side of the molecule. Such an arrangement leads to a better and more effective packing in the crystal lattice of alkanes having even number of carbon atoms. Such a close packing maximises the intersection forces and , therefore, leads to higher melting points for alkanes containing even number of carbon atoms.

4. *Densities :* Alkanes have very low densities. The densities increase gradually in going from lower to higher alkanes, and reaches a limiting value at about 0.8 g/mL (800 kg/m^3). The alkanes are lighter than water. The densities of some alkanes are given in Table 1.7.

Table 1.7 : Densities* of some n-alkanes.

Alkane	*Density* *g/mL*	*kg/m^{-3}*	*Alkane*	*Density* *g/mL*	*kg m^{-3}*
CH_4	0.554	554	H_6H_{14}	0.660	660
C_2H_6	0.564	546	C_7H_{16}	0.684	684
C_3H_8	0.585	585	C_8H_{18}	0.707	707
C_4H_{10}	0.601	601	C_9H_{20}	0.718	718
C_5H_{12}	0.630	630	$C_{10}H_{22}$	0.730	730

* for first four members, the densities are at the respective boiling points, for all others, the densities are at 20°C

5. *Solubility :* Alkanes are non-polar compounds (dipole moment is zero) and hence insoluble in water and other polar solvents. However, they are soluble in non-polar solvents, such as, either benzene, carbon tetrachloride etc. The solubility of alkanes in any solvent, decreases with the increases in molecular mass of alkanes.

GENERAL CHEMICAL PROPERTIES OF ALKANES

- Under ordinary conditions, alkanes are quite inert that is, alkanes do not react with]many regents. This is because,

  (*i*) They do no have any reactive functional group.

  (*ii*) C – C and C – H sigma bonds are strong and require large amount of energy to break.

  (*iii*) C – C and C – H bonds are essentially non-polar and hence electrophilic or nucleophilic substitution reactions do not take place.

However, alkanes undergo substitution reactions in which one or more hydrogen atoms of the alkane molecule are replaced by an equal number of other atoms or groups. The products so formed are termed as the substitution products. Some typical reactions of alkanes are described below.

Substitution Reactions

Alkanes given the following substitution reactions,

(*i*) *Halogenation* : When alkanes react with halogens (chlorine or bromine) in the presence of sunlight or in the presence of a halogen carrier (iron), the H-atoms of alkanes are successively replaced by halogen atoms. For example.

$$\underset{\text{alkane}}{C_nH_{2n+2}} \xrightarrow[-HX]{+X_2} \underset{\text{haloalkane}}{C_nH_{2n+1}X}$$

$$\xrightarrow[-HX]{+X_2} \underset{\text{dihaloalkane}}{C_nH_{2n}X_2} \xrightarrow[-HX]{+X_2} \dots \quad \text{and so on}$$

$$\underset{\text{alkane}}{RH} + X_2 \longrightarrow \underset{\text{haloalkane}}{RX} + HX$$

The order of reactivity is,

F > CI > Br > I

and Tertiary hydrogen > Secondary hydrogen > Primary hydrogen

A carbon atom attached to only one carbon atom is known as primary carbon atom and its three hydrogen atoms are known as primary hydrogen atoms. Similar, carbon atoms attached to two or three carbon atoms are known as secondary and tertiary carbon atoms respectively. The hydrogen atoms attached to these secondary and tertiary carbon atoms are respectively known as secondary and tertiary hydrogen atoms.

In the case of higher alkanes a mixture of all possible isomeric haloalkanes (monohalids) is obtained. The proportional amounts of the isomers formed depend upon the order of reactivity of H atoms.

The effect of halogens on the process of halogenation of alkanes is described below.

(a) *With Fluorine : Fluorine* reacts with alkanes violently and yield hydrogen fluoride and carbon. For example, with methane

$$\underset{\text{methane}}{CH_4} + 2F_2 \longrightarrow C + \underset{\text{hydrogen fluride}}{4HF}$$

However, controlled fluorination in the presence of cobalt moderator, fluoro derivatives are formed.

(b) *With Chlorine* : *Chlorine* in the presence of bright sunlight reacts almost explosively with alkanes, *e.g.*, with methane

$$CH_4 + 2Cl_2 \xrightarrow[\text{sunlight}]{\text{bright}} C + 4HCl$$

However, in the presence of diffused sunlight, hear or halogen carriers, chlorine reacts with alkanes to given chloro derivatives of alkanes. The final product depends upon the amount of chlorine used. For example, when methane is allowed to react with chlorine in the presence of diffused sunlight, the four H atoms of methane are successively replaced by chlorine atoms.

$$\underset{\text{methane}}{CH_4} + Cl_2 \xrightarrow{\text{sunlight}} \underset{\text{chloromethane}}{CH_3Cl} + HCl$$

$$CH_3Cl + Cl_2 \longrightarrow \underset{\text{dichloromethane}}{CHCl_2} + HCl$$

$$CH_2Cl_2 + Cl_2 \longrightarrow \underset{\substack{\text{trichloromeahtne} \\ \text{(chloroform)}}}{CHCl_3} + HCl$$

$$CHCl_3 + Cl_2 \longrightarrow \underset{\substack{\text{tetrachloromethane} \\ \text{(carbon tetrachloride)}}}{CCl_4} + HCl$$

Chlorination of isobutane at 300°C gives a mixture of two isomeric haloalkanes.

$$CH_3-\underset{\substack{| \\ CH_3}}{CH}-CH_3 + Cl_2 \xrightarrow[-HCl]{300^\circ} CH_3-\underset{\substack{| \\ CH_3}}{CH}-CH_2Cl + CH_3-\underset{\substack{| \\ CH_3}}{CCl}-CH_3$$

(c) *With bromine* : Bromine reacts with alkanes in the presence of diffused light/halogen carriers to give bromo derivatives of alkanes similar to chlorine but less readily.

(d) *With iodine* : Iodine reacts with alkanes slowly and reversibly. For example, methane with iodine gives,

$$\underset{\text{methane}}{CH_4 + I_2} \rightleftharpoons \underset{\text{iodomethane}}{CH_3I + HI}$$

Iodination, therefore, is carried out in the presence of an oxidising agent such as, iodic acid (HIO_3), nitric acid or mercuric oxide etc., which oxidise the hydrogen iodide formed during the reaction and permits the reaction to go in the forward direction.

$$\underset{\text{produced in the iodination step}}{5HI} + HIO_3 \longrightarrow 3I_2 + 3H_2O$$

Mechanism of halogenation : Chlorine and bromine react with alkanes through the chain *mechanism*. It involves the following steps.

(a) *Chain initiation step.* In this step, halogen molecule absorbs a photon of suitable radiation from the sunlight (visible region of the electromagnetic radiation) and undergoes homolytic split producing free chlorine atoms.

$$Cl_2 \xrightarrow{hv} \underset{\text{free chlorine atoms}}{2Cl^{\cdot}}$$

or $$CH_3 + {}^*Cl_2 \longrightarrow \underset{\text{chloromethane}}{CH_3Cl} + Cl^*$$

(b) *chain propagation step. The* free halogen atoms produced in step (a) attack the alkane molecule (say CH_4 molecule) to produce hydrogen halide and an alkyl free radical. The alkyl radical so produced is highly reactive and immediately reacts with another halogen molecule to produce haloalkane and releasing free halogen atm. The free halogen atoms so produced then react with another molecule of alkane and the process continuous in a cyclic (chain-like manner).

$$CH_4 + {}^*Cl \longrightarrow \underset{\text{free aradical}}{CH_3} + HCl$$

$$CH_3 + {}^*Cl_2 \longrightarrow \underset{\text{chloromethane}}{CH_3Cl} + Cl^*$$

These reactions proceed to produce more and more of the product. However, depending upon the amount of chlorine in the reaction mixture higher halogen derivatives may also be produced during the halogenation. Thus,

$$\underset{\text{chloromethane}}{CH_3Cl} + \underset{\text{from atom of chlorine}}{Cl} \longrightarrow \underset{\text{chloromethyl radical}}{CH_2Cl} + HCl$$

$$CH_2Cl + Cl_2 \longrightarrow \underset{\text{dichloromethane}}{CH_2Cl_2} + HCl$$

$$CHCl_2 + {}^*Cl \longrightarrow \underset{\text{dichloromethyl radical}}{CHCl_2} \quad HCl$$

$$\cdot CHCl_2 + Cl_2 \longrightarrow \underset{\substack{\text{trichloromethane} \\ \text{(chloroform)}}}{CHCl_3} \quad {}^*Cl$$

$$CHCl_3 + {}^*Cl \longrightarrow \underset{\text{dichloromethyl radical}}{{}^*CCl_3} \quad HCl$$

$$^*CCl_3 + Cl_2 \longrightarrow \underset{\substack{\text{tetrachloromethane} \\ \text{(carbon tetrachloride)}}}{CCl_4} + {}^*Cl$$

(*c*) *Chain termination step.* The above mentioned chain reactions do not continue indefinitely. The chain reaction comes to an end when either the free radicals get deactivated or combine with each other to from a stable (relatively non-reactive) molecule. Such reactions are termed *chain-termination steps.*

In this reaction, the following reactions may lead to the chain termination.

$$Cl^{\cdot} + CH_3 \longrightarrow \underset{\text{chloromethane}}{CH_3CI}$$

$$^{\cdot}CH_3 + {}^{\cdot}CH_3 \longrightarrow \underset{\text{ethane}}{CH_3 - CH_3}$$

$$Cl^{\cdot} + Cl^{\cdot} \longrightarrow Cl_2$$

2. *Nitration :* Alkanes when made to react with nitric acid under suitable conditions give substitution product in which the nitro $(-NO_2)$ group replaces the H-atom in the alkane molecule, viz.,

$$\underset{\text{alkane}}{RH} + HNO_3 \rightarrow \underset{\text{nitroalkane}}{R - NO_2} + H_2O$$

The compounds RNO_2 so obtained are termed **nitroalkanes**. The process of introducing NO_2 group in the molecule in called **nitration**. Higher members can be nitrated by heating with con. HNO_3. The lower members are attacked only by nitric acid vapours at higher temperatures (150 – 475°C) producing a mixture of all possible mononitro derivatives. For example, propane forms a mixture of mononitro derivatives, *e.g.*,

$$\underset{\text{propane}}{CH_3.CH_2CH_3} \xrightarrow[420°C]{NHO_3} \underset{\text{1–nitropropane}}{CH_3.CH_2CH_2NO_2}$$

$$+ \underset{\text{2-nitropropane}}{CH_3\cdot\overset{\overset{NO_2}{|}}{CH}\cdot CH_3} + \underset{\text{nitroethane}}{C_2H_5NO_2} + \underset{\text{nitromethane}}{CH_3NO_2}$$

The liquid-phase nitration of alkanes is slow and generally polynitro compounds are obtained.

Mechanism of nitration : Vapour-phase nitration is believed to proceed by a free-radical mechanism. The details are, however uncertain. A probable mechanism for the nitration of ethane is,

$$CH_3 - CH_3 + NO_2 \longrightarrow CH_3 - CH_2 \quad + HNO_2$$

$$\underset{\text{ethyl free radical}}{C_3 - CH_2} + NO_2 \longrightarrow \underset{\text{nitroethane}}{CH_3 - CH_2NO_2} + CH_3 - CH_2ONO_2$$

$$CH_3 - CH_2ONO \longrightarrow CH_3 - CH_2O + NO$$

$$CH_3 - CH_2O \longrightarrow CH_3 + CH_2O$$

$$CH_3 + NO_2 \longrightarrow \underset{\text{nitromethane}}{CH_3NO_2}$$

(3) *Sulphonation :* Conc H_2SO_4 does not react with alkanes even at elevated temperature. Higher alkanes, from hexane onwards, react with fuming sulphuric acid at high temperature to form alkane sulphonic acids.

$$\underset{\text{alkane}}{RH} + \underset{\text{sulphuric acid}}{HOSO_3H} \xrightarrow{400°} \underset{\text{alkanesulphonic acid}}{R.SO_3H} \quad + H_2O$$

$$\underset{\text{hexane}}{C_6H_{14}} + \underset{\text{oleum}}{H_2SO_4} \xrightarrow{400°} \underset{\text{hexanesulphonic acid}}{C_6H_{13}.SO_3H} \quad + H_2O$$

The sulphonation of n-alkanes is slow, but the branched chain alkanes undergo supohonation readily. The order of reactivity is,

Tertiary hydrogen > Secondary hydrogen > primary hydrogen

Oxidation

Alkanes undergo oxidation yielding products which depend upon the reaction conditions. Some of the oxidation reactions are described below.

(a) *Combustion :* Alkanes are highly combustible and burn readily with a non-luminous flame in excess of air (or oxygen) to give CO_2 and H_2O. The combustions being highly exothermic. It is due to this reaction that the alkanes are used as good fuels (LPG used for cooking purposes is in fact a mixture of lower alkanes). For example,

$$C_nH_{2n+2} + \left(\frac{3n+1}{2}\right) O_2 \longrightarrow nCO_2 + (n+1)\,H_2O + \text{heat}$$

For methane and ethane, the reactions are,

$$CH_4(g) + 2O_2 \longrightarrow CO_2(g) + 2H_2O(g) + 802.5 \text{ kJ}$$

$$2C_2H_6(g) + 7O_2 \longrightarrow 4CO_2(g) + 6H_2O(g) + 2856.5 \text{ kJ}$$

The heat liberated per unit mass of the fuel (the substance undergoing combustion) is termed as the calorific value of the fuel. The commercial value of a fuel is based on its calorific value.

(b) *Slow combustion :* Lower alkanes, such as, methane and ethane when treated with oxygen at high pressure and low temperature undergo slow combustion yielding alcohol, aldehyde, acid and finally carbon dioxide and water. For example, methane gives,

$$\underset{\text{methane}}{CH_4(g)} \xrightarrow{[O]} \underset{\substack{\text{methanol} \\ (\text{methyl alcohol})}}{CH_3OH} \xrightarrow{[O]} \underset{\text{unstable}}{[CH_2(OH)_2]}$$

$$\xrightarrow[-H_2O]{} \underset{\substack{\text{methanal} \\ (\text{formaldehyde})}}{HCHO} \xrightarrow{[O]} \underset{\substack{\text{methanoic acid} \\ (\text{formic acid})}}{HCOOH} \xrightarrow{[O]} CO_2(g)\,H_2O(l)$$

(c) *Catalytic oxidation :* Lower alkanes on oxidation in the presence of a catalyst give alcohols, and aldehydes. Higher alkanes give higher fatty acids. For example, methane gives different products under different conditions.

(i) $$\underset{(9:1 \text{ mixture})}{CH_4(g) + O_2} \xrightarrow[\text{Cu, 200°C}]{\text{100 atm}} \underset{\text{methanol}}{CH_3OH}$$

(ii) $$CH_4(g) + O_2 \xrightarrow[\text{400°C, 200 atm}]{\text{molybdenum oxide}}$$

$$\underset{\text{methanol}}{CH_3OH} \longrightarrow \underset{\text{methanal (formaldehyde)}}{HCHO}$$

(iii) $$\underset{\text{higher alkane}}{2RCH_3} + \underset{\text{from air}}{3O_2} \xrightarrow[\text{100–160°C}]{\text{Mn–acetate}} \underset{\text{fatty acid}}{2RCOOH} + 2H_2O$$

(d) Alkanes are normally resistant towards oxidising agents like $KMnO_4$. However, such oxidising agents oxidise a tertiary hydrogen atom to a hydroxyl group. For example,

$$\underset{\text{isobutane}}{CH_3{-}\overset{\overset{CH_3}{|}}{\underset{\underset{CH_3}{|}}{CH}}} + [O] \longrightarrow \underset{\text{tert. butyl alcohol}}{CH_3{-}\overset{\overset{CH_3}{|}}{\underset{\underset{CH_3}{|}}{C}}{-}PH}$$

Thermal Decomposition

Breaking down of higher alkanes into lower alkanes by heating is termed thermal decomposition. This process is also called as pyrolysis or cracking. In actual practice, the vapours of higher alkanes are passed through a hot metal tube maintained at 500-700°C. For example, n-butane gives.

$$\underset{n\text{-butane}}{CH_3{-}CH_2{-}CH_2{-}CH_3} \xrightarrow[500-700°C]{\text{heat}} \begin{cases} CH_3{-}CH{=}CH{-}CH_3 + H_2 \quad (\text{but-2-ene}) \\ \underset{\text{methane}}{CH_4} + \underset{\text{propane}}{C_3H_6} \\ \underset{\text{ethane}}{C_2H_6} + \underset{\text{ethane}}{C_2H_4} \end{cases}$$

Propane gives,

$$\underset{\text{propane}}{CH_3{-}CH_2{-}CH_3} \xrightarrow{\text{heat}} \begin{cases} \underset{\text{propane}}{C_3H_6} + H_2 \\ \underset{\text{methane}}{CH_4} + \underset{\text{ethane}}{C_2H_4} \\ 2CH_4 + C \end{cases}$$

The process of cracking of great industrial importance. Large quantities of high-boiling fractions (not of much use as such) of petroleum are converted into low-boiling gasoline by cracking.

Isomerisation

The straight chain alkanes when heated with aluminum chloride in the presence of dry HCl at 300°C given branched chain isomers. For example, n hexane gives,

$$\underset{n\text{-hexane}}{CH_3{-}CH_2{-}CH_2{-}CH_2{-}CH_2{-}CH_3} \xrightarrow[300°C]{AlCl_3/HCl} \begin{cases} \underset{\text{2-methylpentane}}{CH_3{-}\overset{\overset{CH_3}{|}}{CH}{-}CH_2{-}CH_2{-}CH_3} \\ \underset{\text{3-methylpentane}}{CH_3{-}CH_2{-}\overset{\overset{CH_3}{|}}{CH}{-}CH_2{-}CH_3} \end{cases}$$

This reaction is of vital importance in petroleum industry. The branched chain alkanes have higher octane number (anti-knock properties).

AROMATISATION

Alkanes containing six or more carbon atoms, when heated under pressure in the presence of suitable catalysts get cyclised to give aromatic compounds. For example, n-hexane give benzene,

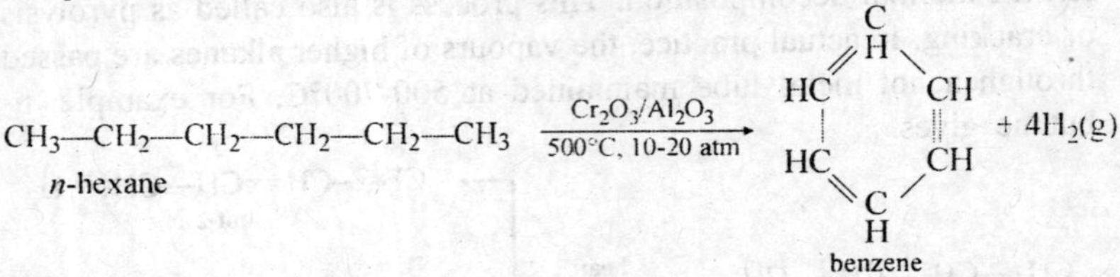

$$\underset{n\text{-hexane}}{CH_3—CH_2—CH_2—CH_2—CH_2—CH_3} \xrightarrow[500°C,\ 10\text{-}20\ atm]{Cr_2O_3/Al_2O_3} \underset{\text{benzene}}{C_6H_6} + 4H_2(g)$$

n-heptane gives toluene,

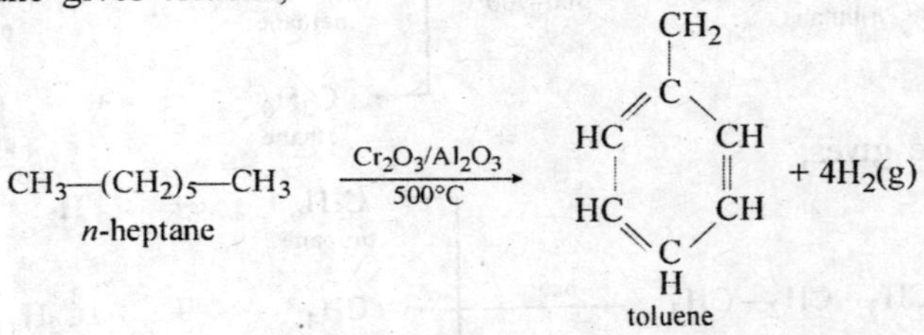

$$\underset{n\text{-heptane}}{CH_3—(CH_2)_5—CH_3} \xrightarrow[500°C]{Cr_2O_3/Al_2O_3} \underset{\text{toluene}}{C_6H_5CH_3} + 4H_2(g)$$

Fig. 1.10

Uses of Alkanes

Alkanes are the simplest organic compounds containing carbon and hydrogen. Some major uses of these compounds are,

(i) Lower alkanes occurring as natural gas and lighter petroleum fractions are used as fuels.

(ii) Low-boiling liquid alkanes *e.g.*, hexane, are used as solvents.

(iii) Heavy petroleum fractions are used as lubricants (grease), and for obtaining waxes and vaseline.

(iv) The products of cracking process are generally used for producing linearalkyl benzenes, (LAB) used as the raw material for manufacturing detergents.

CONFORMATIONS OF ALKANES

The carbon-carbon single bond (sigma bond) in alkanes is cylindrically symmetrical about the internuclear axis (line joining the two carbon nuclei). *The symmetry of the bond permits free rotation of one methyl group relative to others about the bond axis without breaking the bond.* The rotation about a sigma bond is not completely free. In fact, it is hindered by an energy barrier of 1–20 kJ/mol. The repulsive interaction between the neighboring bounding electron pairs is called *torsional strain.* As a result, a number of different spatial arrangements of the atoms or groups attached to the carbon atoms are possible.

The different structures (arrangements) which can be produced by the rotation about the carbon-carbon single bond are called *conformers.* The molecular geometry corresponding to a conformer is known as *conformation.*

Conformers are usually represented using either the *Sawhorse projection* or the *newman projection.*

Conformations of Ethane

In ethane ($CH_3 - CH_3$) molecule, the two carbon atoms are bounded together by a single bond. Due to the free rotation about the carbon-carbon bond an infinite number of arrangements of hydrogen atoms attached to one carbon atom with respect to the hydrogen atoms on the other carbon atom are possible. Thus, in ethane an infinite number of conformations are possible. However, only two conformations in between these two extreme conformations are called *gauche* or *skew*. These are shown in Fig. 15.4, and described below.

(i) *Eclipsed from :* The arrangement in which the hydrogen atoms of one carbon lie exactly behind those of the other is called eclipsed from.

(ii) *Staggered from :* The arrangement in which the hydrogen atoms of the nearer methyl group are seen exactly between those of the other methyl group, *i.e.*, the hydrogen atoms are staggered with respect to each other, is called staggered from.

These conformations are represented by Sawhorse, and Newman projections.

The Sawhorse projections of the eclipsed and the staggered forms of ethane are shown in Fig. 1.11.

Newman projection of the conformations : Newman projection is easier to draw and visualise. To convert a conformer from a Sawhorse (or structural) from to the Newman from, the following procedure is adopted.

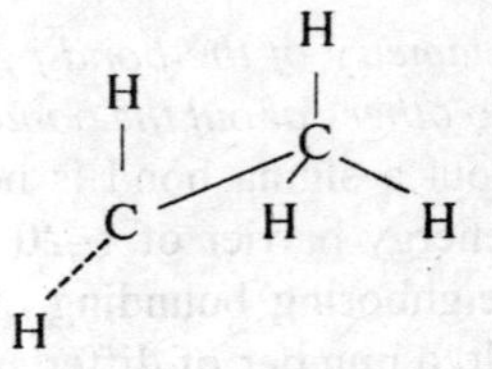

Eclipsed form of ethane molecule

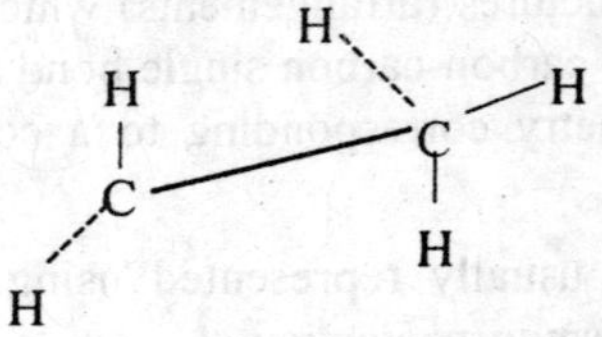

Staggered form of ethane molecule

Fig. 1.11 : The Sawhorse projections of the confformations of ethane.

(a) The two carbons bonded by a single bond are represented by a small circle, the front carbon covering the second carbon lying exactly behind it.

(b) The C—H bonds of the front (nearer) carbon are shown by solid lines starting from the centre of the circle.

(c) The C—H bonds of the real carbon (carbon behind the front carbon) are shown by dotted lines starting from the circumference of the circle.

The Newman projection for the eclipsed and the staggered forms of ethane molecule are shown in Fig. 1.12.

Eclipsed Staggered Skew

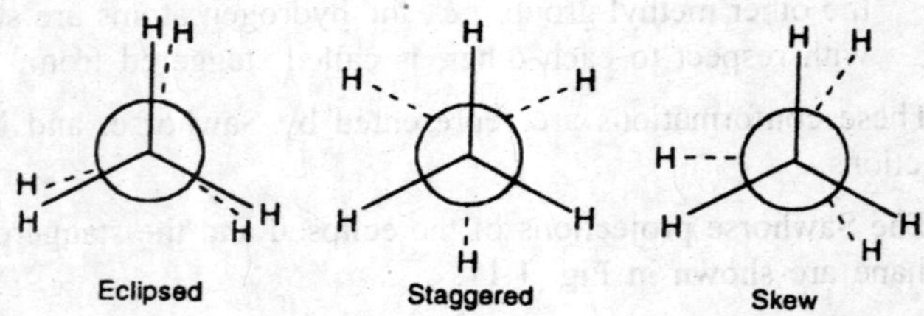

Fig. 1.12 : Eclipsed and Staggered conformations of ethane.

Relative Stabilities of Eclipsed and Staggered Conformations of Ethane

The various conformations of a compound posses identical physical characteristics but differ in energy.

In the eclipsed conformation, all the C—H bonds of one methyl group are aligned with those of the second, and in the staggered from the two methyl groups are rotated as far away as possible from one another.

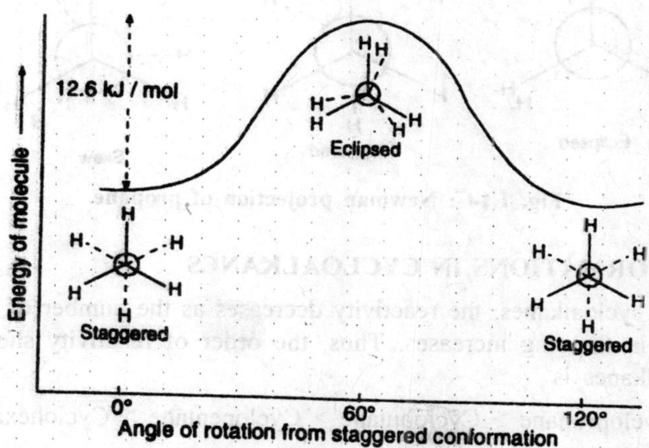

Fig. 1.13 : Degree of rotation from the Staggered form and energy of the molecule of ethane.

Thus, the H-atoms on the two carbon atoms are closer in the eclipsed from than in the staggered from. So , the repulsive force acting between the two hydrogen atoms in eclipsed from is more than that in the staggered form.

As a result, the energy of the eclipsed form is higher than that of the staggered form of ethane. Thus, the staggered conformation of ethane is more stable than the eclipsed form. The order of stability for different conformations is

Staggered > Skew (or Gauche) > Eclipsed

The staggered conformer of ethane is more stable than the eclipsed conformer by about 12.6 kJ mol^{-1}. As the energy difference is small, there is no effective energy barrier to the rotation about a sigma bond.

Thus, we can say that in ethane, there is free rotation about a C—C sigma (s) bond.

At room temperature, the eclipsed and staggered forms of ethane interconvert rapidly and cannot be isolated as separate conformers.

Conformation of Propane

Eclipsed Staggered Skew

Fig. 1.14 : Newman projection of propane.

CONFORMATIONS IN CYCLOALKANES

In cycloalkanes, the reactivity decreases as the number of carbon atoms in the ring increases. Thus, the order of reactivity shown by cycloalkanes is

Cyclopropane > Cyclobutane > Cyclopentane > Cyclohexane

A. Von Baeyer explained the decreasing reactivity of cycloalkanes with increasing number of carbon atoms in the ring in terms of deviation of the $C\hat{C}C$ angle in any cycloalkane from the tetrahedral angle of 109°28. According to *Baeyer's Strain, Theory, any deviation of bond angle from the normal tetrahedral angle of 109°28′ would cause an internal strain in the ring. The larger the angle of deviation, larger the strain and hence greater is the reactivity of the molecule.*

Assuming all cycloalkane to be planar, the angle strain in various cycloalkanes are,

Cycloalkane:	Cyclopropane	Cyclobutane	Cyclopentane	Cyclohexane	Cycloheptane
Angle strain:	+ 24°44′	+ 9°44′	+ 44′	–5°16′	–9°33′

The + sign indicates that the $C\hat{C}C$ angle is compressed to satisfy the geometry of the ring. The –ve sign indicates widening of the $C\hat{C}C$ angle to satisfy the geometry of the ring. *The strain in the molecule and, therefore, its reactivity is determined by the magnitude of the angle strain irrespective of its sign.*

From the angle strain values given above, we would expect cyclopropane to the most reactive of all the cycloalkanes listed, and cyclopentane to be the least reactive. This is found to be so. However, from the angle strain values, cyclohexane and higher cycloalkanes should show greater reactivity than cyclopentane. But, cyclohexane and higher cycloalkanes are found to be quite stable.

Conformations in cyclohexane : To explain the stability (or reactivity) of higher cycloalkanes, Sachse and Mohr (1998) pointed out that the rings of cyclohexane and higher cycloalkanes can become free of strain if all the carbon atoms in the ring are not forced into one plane. Thus, the molecules of cyclohexane and higher cycloalkanes avoid the angle strain by adopting conformations in which all the bond angles are close to 10°28' and are therefore, quite stable and unreactive. The two most important conformations for cyclohexane are the *chair* and the *boat forms* shown in Fig. 1.15. The Newman projections of cyclohexane conformers are shown in Fig. 5.16.

(*a*) chair conformer

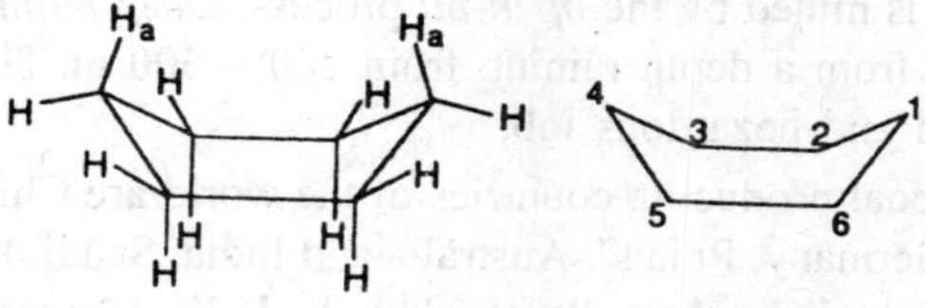

(*b*) boat conformer

Fig. 1.15 : Chair and boat conformation of cyclohexane.

chair conformer

boat conformer

Fig. 1.16 : Newmann projection of cyclohexane

Of the two, the chair form is more stable than the boat form, the energy difference being about 44kJ mol^{-1}, *i.e.*, the boat form has 44kJ/mol more energy than the chair form of cyclohexane. Therefore, *cyclohexane exists mainly in the chair form*. This is because in the boat form, many hydrogens on the adjacent carbons correspond to the unfavourable eclipsed conformation of ethane. Also, the two hydrogens marked Ha (commonly called flag pole hydrogens) in the boat form are quite close and repel one another. On the other hand, the chair form does not have these unfavourable interactions and all the hydrogens correspond to the more stable staggered conformation of ethane.

COAL AND PETROLEUM

Until the early years of the nineteenth century, only the plants and animals were the known sources of organic compounds. With the advent of the Industrial Revolution in Europe, fossil fuels like coal and petroleum gained prominence as the sources of organic compounds, particularly hydrocarbons. Today, hydrocarbons are mostly obtained from petroleum. However, in recent years coal has started gaining prominence, probably due to the uncertain conditions in the world oil market.

Coal and petroleum have organic (plant and animal) origin.

Coal

Coal is one of the earliest used fossil fuels. It occurs at varying depths under the upper crust of the earth. When found at low depths (20 – 50 m), it is mined by the *open-pit* process. *Deep mining* involves mining of coal from a depth raining from 200 – 500 m. Deep mining is a specialized and hazardous job.

The chief coal producing countries of the world are China, U.S.A., Russia, U.K., Germany, Poland, Australia and India. Saudi Arabia, Iraq, the countries in the West Asia, Russia, U.S.A., U.K., Mexico and China are important oil producing countries. In India, coal is mainly mined in Bihar. West Bengal, Madhya Pradesh and to a lesser extent in Andhra Pradesh.

Composition of Coal

Coal is mainly composed of carbon. Depending upon the carbon content of coal, it is classified into different varieties. Common varieties of coal are,

Variety of coal:	Peat	Lignite	Bituminous	Anthracite
Carbon content:	10-15%	40%	60-70%	80-90%

Coal also contains arenes (aromatic hydrocarbons) like benzene, toluene, xylene, naphthalene and anthracene besides some organic compounds of sulphur and nitrogen.

Formation of Coal in Nature

It is believed that coal in nature was formed from the remains of the trees buried inside the earth some 500 million years ago. Due to the bacterial and chemical action on the wood, it got converted into **peat.** Then, as a result of high temperature and high pressure inside the earth peat got transformed into coal.

$$\text{(dead plants and trees)} \xrightarrow[\text{and chemical action}]{\text{bacterial}} \text{Peat} \xrightarrow[\text{pressure}]{\Delta} \text{Lignite} \xrightarrow[\text{pressure}]{\Delta} \text{Bituminous coal} \xrightarrow[\text{pressure}]{\Delta} \text{Anthracite}$$

Aromatic Compounds From Coal

Coal is a complex mixture of hydrocarbons. It also contains some organic compounds containing nitrogen and sulphur in small amounts. It can be approximated to the formula $(C_3H_4)_n$

Coal is hydrogen-deficient substance. It is assumed that the basic structure of coal is probably built up of a large number of interlocked benzene rings, upto thirty in bituminous/anthracite coal. Hydrogen is present in the aliphatic sidechains. Approximate composition of some common varieties of coal are given below.

Table 1.8 : Approximate composition of some common varieties of coal

Element	*Peat (Yellow to light brown)*	*Lignite (Black)*	*Bituminous (Bright black)*	*Anthracite (Brown)*
Carbon	54.0	64.0	84.0	93.0
Hydrogen	5.5	5.0	4.5	3.3
Oxygen	35.0	26.0	8.0	2.0
Nitrogen	2.0	1.3	1.3	1.3
Sulphur	3.5	3.3	1.5	0.5

For obtaining organic compounds from coal, bituminous coal is thermally decomposed. This process is called destructive distillation of coal or pyrolysis of coal.

Destructive Distillation of Coal

Degradation (breakdown) of coal at high temperature in the absence of air is termed destructive distillation of coal. For this purpose, coal is heated in iron retorts in the absence of air, when the volatile material evolved is collected as distillate/condensate.

The process can be carried out under different temperature conditions, depending upon the nature of the products required.

Low Temperature Pyrolysis

Coal when heated in the absence of air at bout 723-973 K, gives soft solid, smokeless coke called *coalite*. The yield of coat tar and coal gas is doubled. Coal tar so obtained contains more of aliphatic compounds.

High Temperature Pyrolysis

In this process coal is heated in the absence of air upto about 1273-1473 K. The major products of high temperature pyrolysis are,

The yield per ton of coal on destructive distillation at higher temperature is shown in Fig. 1.17.

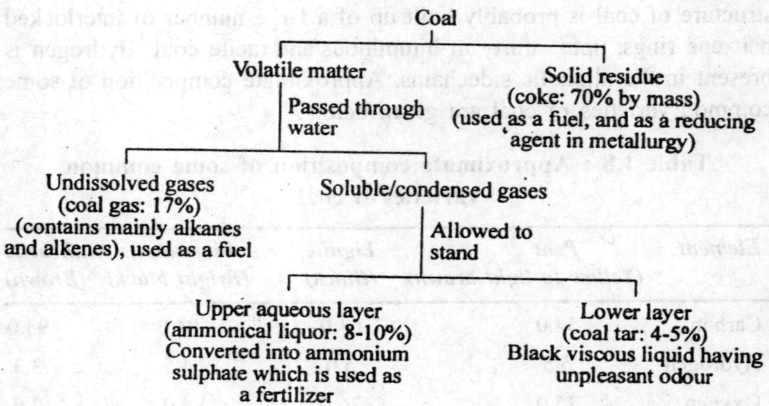

Fig. 1.17

Coke	Coal gas	Coal tar	Ammonium sulphate	Light oil
680 kg	30m^3	15-22 kg	1 kg	5-6 kg

Coal tar obtained as a heavy viscous dark* liquid with an unpleasant odour is a valuable by-product in the destructive distillation of coal. Its composition depends upon the nature of coal used during distillation. However, coal tar generally contains the following compounds

Acid compounds.	Phenol and cresols etc.
Basic compounds.	Pyridine, etc.
Neutral compounds.	Benzene, toluene, naphthalene and anthracene etc.

FRACTIONAL DISTILLATION OF COAL TAR

Coal-tar obtained from the destructive distillation of coal is preheated in a preheater heated by the outgoing hot gases. Here, coal tar loses most of the water it contains, along with some low boiling hydrocarbons.

Then, it is sent to the main iron still (a side retort) heated directly in a furnace. The vapours are condensed to obtain various fractional distillation of coal-tar are given below in Table 1.9.

Table 1.9 : Various fractionals obtained from the fractional distillation of coal tar

Fraction	*Temperature range* °C	*K*	*Name of fraction*	*Constituents*	*Percentage (V/V)*
1.	upto–170	upto–443	Light oil	Benzene, Toluene, Xylene	2.25
2.	170–230	443-503	Middle oil	Phenol, Cresol, Naphthalene	7.5
3.	230-270	503-543	Heavy oil (Creosote oil)	Cresols, Naphthalene, Naphthol	16.5
4.	270-360	543-533	Antracene oil (Green oil)	Anthracene	12
5.	–	–	Pitch (residue in the still)	Carbon	56

* The black colour of coal tar is due to presence of carbon particles in it.

PETROLEUM

Petroleum occurs deep inside the earth at a depth ranging from a few hundred to few thousand meters, Fig. 1.18. The name petroleum has been given to the crude oil because it occurs under the earth's crust entrapped in rocks (petra = oleum = oil). Crude oil (petroleum) is pumped out of earth by drilling oil wells.

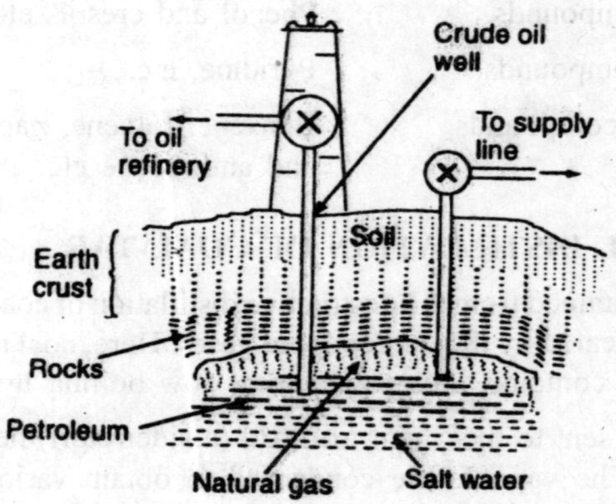

Fig. 1.18 : Occurrence of petoleum under the earth crust.

Type chief petroleum producing countries are, U.S.A., Russia, Saudi Arabia, Kuwait, Iraq, Libya, Nigeria, Algeria, Venezuela and Mexico. In India, oil fields are located in Assam., Gujarat, and Maharashtra. For the last few years, we have started producing oil from the high seas off Mumbai.

Petroleum is obtained by drilling a hole through the crust till it reaches the oil reserve. The oil gushes out of its own due to high pressure of the gas over its surface. However, when the pressure falls, it is pumped out with the help of pumps.

Large quantity of natural gas is generally associated with petroleum. Petroleum is often referred to as *liquid gold*, due to its importance in commerce and industry. Prosperity of any country depends upon its petroleum reserves.

COMPOSITION OF PETROLEUM

Crude oil (or petroleum) is a dark brown (sometimes with greenish tinge) viscous liquid. This crude oil is a mixture of about 150 different organic compounds.

Crude oil is largely made up of aliphatic hydrocarbons with much lesser amounts of aromatic compounds and organic compounds of sulphur and nitrogen.

FORMATION OF PETROLEUM IN NATURE

It is believed that *petroleum (or crude oil) was formed due to the decomposition of the prehistoric vegetable micro-organisms and sea animals under high temperature, high pressure and in the absence of the air.*

The remains of the dead micro-organisms sank to the bottom of the sea. Gradually, these were covered by sand and clay. Over a period of millions of years, these remains got converted into petroleum under the combined effect of high temperature, high pressure and bacteria in the absence of air. The petroleum so formed passed through porous rocks until it got trapped between some impervious (nonporous) rocks.

The more volatile components of petroleum from a gaseous cover petroleum. Such a gaseous mixture covering the crude oil is known as *nature gas.*

HYDROCARBONS FROM PETROLEUM

The crude oil or petroleum as pumped out of the oil well is a viscous and complex mixture of several hydrocarbons and small amounts of other compounds. Crude oil (petroleum) cannot be used as such. Crude petroleum, however, can be separated into many useful fractions by fractional distillation. The process of separating crude oil into useful fractionals is called refining of petroleum.

In India, oil refineries are located at Koyali, Mumbai, Chennai, Kochi, Visakhapatnam,, Haldia, Barauni, Guwahati, Digboi and Mathura.

Refining of Crude Oil

The crude oil is first washed with acidic or basic solution as needed. Then, the crude oil is heated in a France to 650 K and the pressure reduced.

The vapours of the crude oil are then fed into a fractionating column through an inlet near its bottom. Fractionating tower is a tall cylindrical steel structure with horizontal trays fitted in it. Each tray has a number of holes and a short tube with a bubble cap. High boiling fraction condenses in the lower section of the tower, while the vapours move up. Due to a regular temperature gradient along the height of the column, the fraction with different boiling points get liquefied at different heights, and can be tapped out (Fig. 1.19).

Major products obtained from the fractional distillation of crude oil are listed.

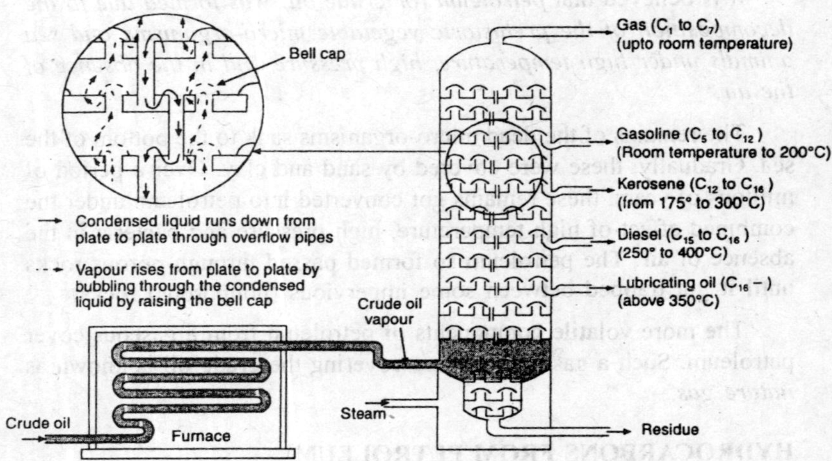

Fig. 1.19 : Fractional distillation of crude oil. The cutway view shows how ractionation of nore volatile components takes place.

CRACKING OF HYDROCARBONS

The process by which higher hydrocarbons having high boiling points are broken into a variety of lower hydrocarbons, which are more volatile (low boiling), is called **cracking** (or sometimes as **pyrolysis).** For example, a higher hydrocarbon $C_{10}H_{22}$ splits according to the reaction

$$\underset{\text{higher hydrocarbon}}{C_{10}H_{22}} \xrightarrow[770\text{ K}]{\text{heat}} \underset{\text{lower hydrocarbons}}{C_8H_{18} + C_2H_4}$$

The process of cracking, thus i cracking, carbon-carbon bonds get its are formed during cracking, carbon-carbon bonds get broken in random manner. So various kinds of products are formed during cracking.

Creaking is generally carried out in the following ways.

Thermal cracking : Breaking down large molecules by heating at high temperature and pressure termed thermal cracking. Thermal cracking can be further classified into the following classes.

(*a*) *Liquid phase thermal cracking* : In this process, the higher boiling lubricating oil are converted into low boiling fractions by temperature of 750 K, and under a pressure of about 10

(*b*) *Vapour phase thermal cracking* : In this process, low boiling fraction, *e.g.*, kerosene is cracked in the vapour phase at a temperature of about 875 K and under a atmosphere.

(*i*) *Catalytic cracking* : Higher hydrocarbons can be cracked at lower (600-650 K) and lower pressure (2 atm.) in the presence of a suitable catalyst. A typical catalyst used for this purpose is a mixture of silica (SiO_2), 4 parts; alumina (Al_2O_3), 1 part, and manganese dioxide (MnO_2), 1 part. Catalytic cracking produces gasoline of higher octane number and, therefore, this method is used for obtaining better quality gasoline.

(*ii*) *Steam cracking* : In this process, higher hydrocarbons in the vapour phase are mixed steam, heated for a short duration to about 900°C, and cooled rapidly. The process is suitable for obtaining lower unsaturated hydrocarbons.

APPLICATIONS OF CRACKING

Of the various products obtained hi straight-run refining, the most important are petrol, diesel and kerosene. The demand for these products is more than what is obtained during refining of the petroleum. On the other hand, high-boiling fractions find lesser use as such. So,

(i) In petroleum industry, the yield of the low boiling (lower molecular mass) fractions, such as gasoline is increased by the cracking of less useful high boiling fractions.

For example, large molecules hi the lubricating oil range (C_{18} to C_{21}) can be converted into more useful molecules in the petrol (gasoline) range (C_6 to C_9) by cracking process.

$$\underset{\text{hexadecane}}{C_{16}H_{34}} \xrightarrow{\text{cracking}} \underset{\substack{\text{octane} \\ \text{(alkane)}}}{C_8H_{18}} + \underset{\substack{\text{oct-1-ene} \\ \text{(alkane)}}}{C_8H_{16}}$$

(ii) Cracking always yields low boiling alkenes as the by-products. These unsaturated hydrocarbons called petrochemicals, are converted into a variety of useful compounds such as, polyethylene etc.

REFORMING, OR AROMATISATION

The process of converting open chain (aliphatic) hydrocarbons and/ or cycloalkanes into aromatic hydrocarbons (arenes) containing the same number of carbon atoms in the presence of a catalyst is called **reforming** or **aromatisation.** Reforming (or aromatisation) involves reactions of the type, n, cyclisation, and isomerisation.

In reforming (or aromatisation), cyclic and acyclic alkanes containing six to eight| atoms are heated at about 670 K in the presence of palladium, platinum or nickel as i Platinum seems to *be* the best catalyst and so the process is sometimes called **platforming.**

Following are the two examples of reforming.

Hexane when passed over Cr_2O_3 supported over alumina at 670 K undergoes cyclization and finally to aromatisation, forming cyclohexane and benzene respectively.

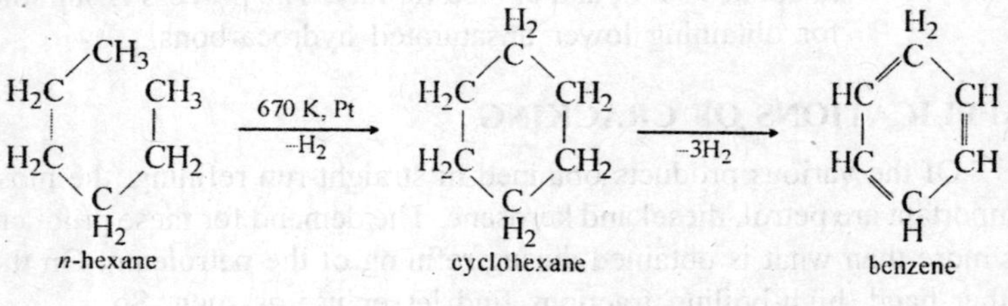

n-hexane —(670 K, Pt; $-H_2$)→ cyclohexane —($-3H_2$)→ benzene

Fig. 1.20

Under similar conditions, *n*-heptane yields methylcyclohexane and finally toluene.

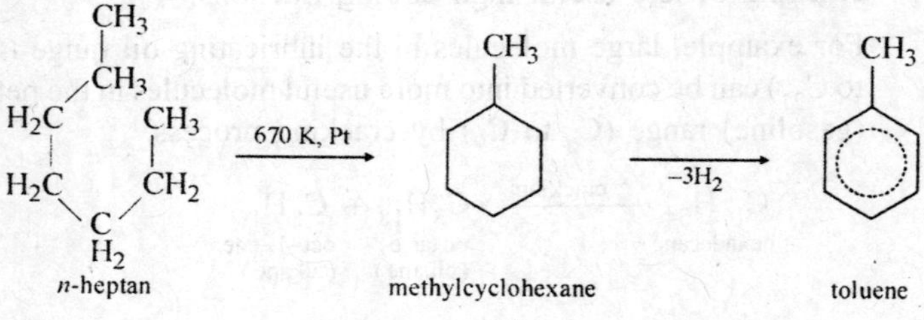

n-heptan —(670 K, Pt)→ methylcyclohexane —($-3H_2$)→ toluene

Fig. 1.21

APPLICATIONS OF REFORMING

The process of reforming has been employed for,

(i) Producing aromatic hydrocarbons (arenes) like benzene, toluene and xylenes. Prior to the development of petroleum sources, the bulk of these arenes were obtained from coal tar. But now they are derived from petroleum by the process of reforming.

(ii) Increasing the proportion of aromatic hydrocarbons in gasoline to increase the octane number of straight-run gasoline. Aromatic hydrocarbons have octane numbers higher than 100, while the straight-run gasoline has the octane number of 55.

The straight-run gasoline (octane number 55) is first distilled to obtain a fraction containing n-hexane, cyclohexane and methylcyclopentane. This fraction is then heated with a platinum catalyst to convert the cyclohexane into benzene and toluene (octane number > 100).

FLASH POINT OF A HYDROCARBON

The explosive nature of a hydrocarbon is determined by its volatility. The volatility of any liquid depends upon the temperature.

"The lowest temperature at which a liquid hydrocarbon gives off enough vapour to form an explosive mixture with air is termed as its flash point."

The flash point of any liquid hydrocarbon is so adjusted that it remains safe under the conditions of its use. For example, the minimum flash point permitted in India is 44°C, while in France it is 33 °C and in Britain it is 22°C.

Determination of Flash Point

Flash point of a combustible liquid is determined by using Abel's apparatus shown in Fig. 1.22.

The given sample of the combustible liquid, *e.g.*, kerosene, petrol, diesel etc., is taken in a small metal cup and heated slowly through a surrounding water bath. From time to time, the slit in the cover of the metal cup is opened, and a tiny flame introduced momentarily by means of an automatic device. The temperature at which the vapours get ignited with a minute explosion is taken as the flash point of the liquid.

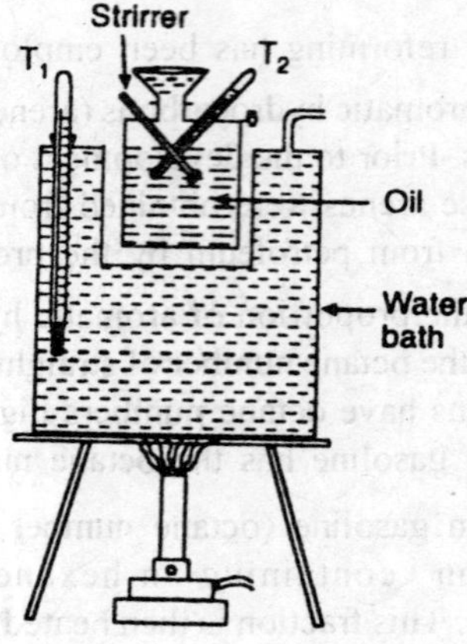

Fig. 1.22

KNOCKING OF FUELS

An internal combustion engine works with a system of pistons^ In the down-stroke of the piston, a mixture of air and gasoline vapour is drawn from the carburetor into the cylinder, and in the upstroke, the mixture is compressed. The ratio of the initial volume to the final volume is called the compression ratio. At the end of the upstroke of the piston, a spark ignites the compressed air-gasoline mixture. As the gases bum, they expand and the flame front moves in a smooth manner and supplies power to the engine.

For maximum efficiency of the engine, a high compression ratio (generally seven to eight) is required. But, as the compression ratio increases, the gasoline-air mixture burns in an explosive manner and produces a metallic sound. This sound is called knocking, and indicates inefficient performance of the gasoline. High compression ratio engines also require less fuel, so gasoline having less knocking tendencies are very valuable.

It has been found that the knocking tendency of the fuels falls off with the nature of the fuel as follows.

Straight chain alkanes > Branched chain alkanes > Alkenes

Knocking may also be prevented or minimised by adding compounds such as tetraethyl lead (TEL) to gasoline. Such compounds are called anti-knocking agents. Due to toxic effects of lead, the use of TEL in petrol has been prohibited. These days, the knocking behavior of petrol (gasoline) is improved by reforming process (cyclisation, aromatisation, isomerisation) or by adding iso compounds.

QUALITY OF PETROL (GASOLINE)–OCTANE NUMBER

An internal combustion engine gets its power by controlled combustion of fuel in the presence of air. If a low quality fuel is used, the fuel does not burn smoothly. An explosive sound is produced occasionally. It is known as knocking. This greatly reduces the power of the engine.

Different hydrocarbons have different knocking tendencies. A fuel which produces minimum knocking is considered as a good fuel. *The quality of a fuel is indicated in terms of its octane number.*

. An arbitrary scale of octane number has been set up with n-heptane and 2,2,4-trimethylpentane (iso-octane) as the reference compounds. 2,2,4-trimethylpentane (generally called iso-octane) has excellent anti-knocking properties. So, iso-octane was arbitrarily assigned an octane number of 100, whereas n-heptane which is very prone to knocking was assigned an octane number of zero (0).

$$CH_3—CH_2—CH_2—CH_2—CH_2—CH_2—CH_3$$

n-hep tan e
Oc tan enumbe . zero

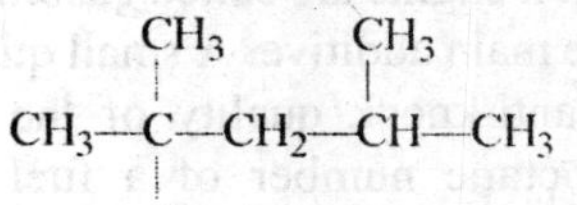

$$CH_3—C(CH_3)_2—CH_2—CH(CH_3)—CH_3$$

iso-octane (2,2,4-trimethylpentane)
Octane number: 100

Fig. 1.23

All fuels are graded in between these two limiting values by comparing with a suitable mixture of the above two compounds. Thus, *the octane number of any fuel is defined as the percentage of iso-octane in a mixture of iso-octane and n-heptane that has the same knocking as the fuel under examination.* Thus, a fuel having an octane number of 80 behaves in a i mixture having 80% of iso-butane, and 20% of n-heptane. Straight run gasolines may have octane values ranging from 20 to 73. Aviation fuel is rated as 100 octane.

Therefore, the antiknock property of a fuel increases with the increase in its i Branched-chain alkanes, alkenes and aromatic hydrocarbons have higher compounds having higher octane numbers are added to straight-run gasoline to raise Its octane number. These high octane compounds, viz., branched-chain alkanes, alkenes, and hydrocarbons are prepared by catalytic cracking and catalytic reforming. *The octane number* decreases *with an increase in chain length.*

Octane ratings of some compounds are given below.

n-Heptane	0	Methanol	116
n-Pentane	62	Toluene	118
tert-butyl alcohol	98	Methane	122
iso-octane	100	Ethane	101
Benzene	106	Propane	96
Ethanol	112	Butane	89

Recently, the hydrocarbon, 2,2,3-trimethylpentane, called **triptane** has been assigned an octane number 125.

GASOLINE ADDITIVES

Certain compounds added to gasoline for improving its combustion in an internal combustion engine are called *gasoline additives*. Tetraethyl lead ($(C_2H_5)_4Pb$) is the main additive. A small quantity of tetraethyl lead (TEL) improves the anti-knock quality of the gasoline. Compounds which improve the octane number of a fuel are called *antiknock compounds*. Tetraethyl lead (TEL, $(C_2H_5)_4Pb$) is a common antiknock compound. Gasoline containing tetraethyl lead is called *leaded* gasoline or ethyl gasoline, or simply as *leaded petrol*.

A gasoline-engine (internal combustion engine) discharges various substances into the atmosphere. These are carbon dioxide, water, carbon monoxide, nitrogen oxides, unburnt hydrocarbons and from leaded gasoline certain compounds of lead. About 0.1 g of lead per litre of the gasoline burnt escapes into the atmosphere. Most of it finds its way into the soil. Although many compounds present in the exhaust of any automobile are harmful, lead is extremely toxic and poisonous. The increasing emission of such harmful substances into the atmosphere due to the increasing use of automobiles, is a serious pollution problem. Certain devices such as, catalytic converters are now being used to oxidise the unburnt hydrocarbons and carbon monoxide to carbon dioxide and to decompose nitrogen oxides into nitrogen and oxygen.

An important step towards pollution control involves removal of lead from gasoline. This is done by replacing tetraethyl lead in gasoline by certain aromatics and branched-chain hydrocarbons. The gasoline free from tetraethyl lead is called **lead-free gasoline** (or lead-free petrol).

QUALITY OF DIESEL–CETANE NUMBER

The quality of diesel is expressed in terms of its **cetane number.** Cetane number is based on a scale in which

Hexadecane ($C_{16}H_{34}$) called cetane, is given a cetane number of 100

1-methylnaphthalene (C_2H_{10}) is given a cetane number of 0.

The cetane number is defined as the percentage of hexadecane (cetane) by volume in a mixture of hexadecane and 1-methylnaphthalene, which has the same ignition quality as the fuel under testing.

SYNTHETIC PETROL (OR GASOLINE)

Mixture of alkanes resembling petroleum (crude oil) can be obtained from coal by two methods, namely the Fischer-Tropsch process and the Bergius process.

These processes were developed in Germany during World War II, when is petroleum supplies were cut off. Germany produced considerable amounts of fuel from coal by the above processes during that period.

These process have once again drawn much attention due to the existing uncertainties in the world oil markets.

Bergius process : In this process, powdered coal is mixed with heavy oil and heated with hydrogen under high pressure (200-250 atm) and at about 748 K in presence of iron oxide as catalyst.

$$\underset{\text{from coal}}{nC} + (n+1)\,H_2\,(g) \xrightarrow[200-250\text{ atm},\,748\text{ K}]{\text{iron oxide}} \underset{(\text{mixture of hydrocarbons})}{C_nH_{2+2}}$$

The vapours on condensation give a liquid resembling crude oil. This is called synthetic petroleum, which on fractional distillation gives petrol (gasoline).

Fisher-Tropsch process : In this process, a mixture of water gas and hydrogen under pressure (5 = 10 atm) is passed over a cobalt catalyst at 450-475 K. The water gas required is obtained by passing steam over red hot coke.

$$\text{C (red hot)} + H_2(g) \rightarrow \underset{\text{water gas}}{CO + H_2}$$

$$nCO + (2n+1)\,H_2 \xrightarrow[\text{5–10 atm, 450–475 K}]{\text{Co catalyst}} C_nH_{2n+2} + nH_2O$$

The petroleum-like product obtained in the process in fractionally distilled to obtain petrol, middle oil and heavy oil. The middle oil fraction is hydrogenated again to produce more petrol.

PETROCHEMICALS AND THEIR APPLICATIONS

Petroleum is a mixture of a large number of saturated hydrocarbons. Many lower saturated and unsaturated hydrocarbons can be produced by cracking process. These lower saturated and unsaturated hydrocarbons are very important raw materials for many useful organic compounds.

Various low molecular mass organic compounds prepared from the lower hydrocarbons obtained from natural gas of/and petroleum, are termed petrochemicals.

Examples:

Some important petrochemicals are,

(i) Methane (CH_4)

(ii) Methanol (methyl alcohol, CH_3OH)

(iii) Ethanol (ethyl alcohol, C_2H_5OH)

(vi) Ethene (ethylene, $CH_2 = CH_2$)

(v) Propene (propylene, $CH_3 - CH = CH_2$)

(vi) Butene (butylene, C_4H_8

(vii) Butadiene (C_4H_6)

(viii) Benzene (C_6H_6)

(ix) Toluene ($C_6H_5CH_3$)

(x) Xylene (C_6H_4 $(CH_3)_2$)

INDUSTRIAL USES OF PETROCHEMICALS

All petrochemicals are very good raw materials for the preparation of a large number of useful organic compounds such as,

(i) Fertilizers (*e.g.*, urea)

(ii) Insecticides (*e.g.*, DDT, BHC etc).

(iii) Synthetic detergents (*e.g.*, linear alkyl benzenesulphonate)

(iv) Synthetic polymers (*e.g.*, PVC, polythene, polypropene etc).

(v) Synthetic fibres (*e.g.*, polyester)

(vi) Synthetic rubbers (*e.g.*, neorpene)

(viii) Dyes, varnishes and drugs etc.

PRODUCTION OF PETROCHEMICALS

Petrochemicals are obtained from various fractions of petroleum by using process such as, cracking, cyclisation, oxidation, polymerisation, halogenation etc.

Major hydrocarbon petrochemical obtained from petroleum refineries are shown, below.

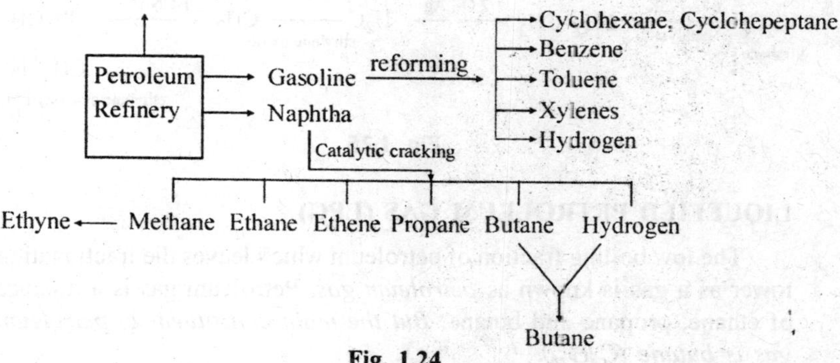

Fig. 1.24

SOME INDUSTRIALLY IMPORTANT CHEMICALS FROM METHANE AND ETHANE

Some important chemicals that act as the raw material from manufacturing many industrially important products are obtained from methane and ethane as shown in the following flow sheet charts.

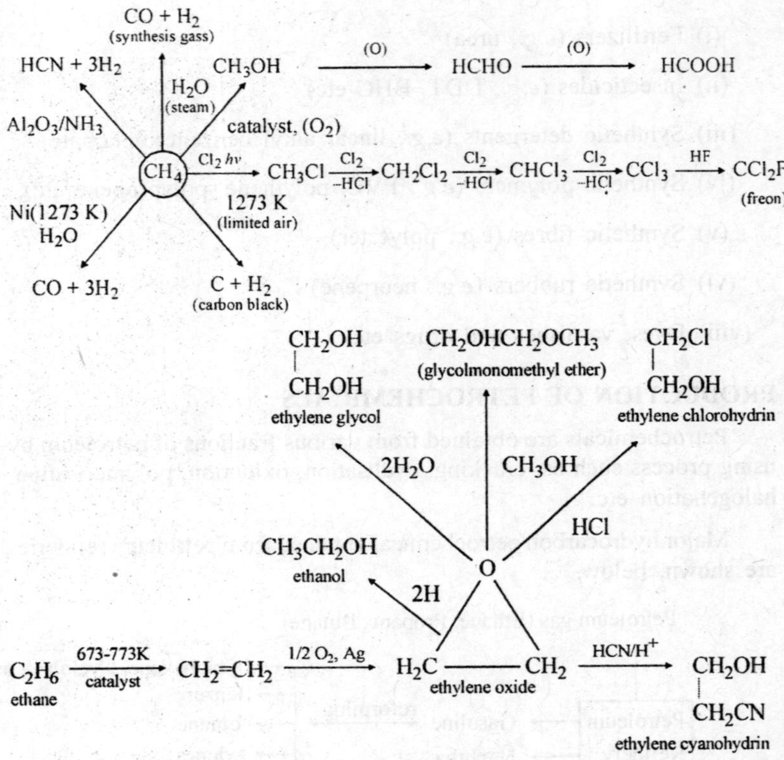

Fig. 1.25

LIQUEFIED PETROLEUM GAS (LPG)

The low boiling fraction of petroleum which leaves the fractionating tower as a gas is known as *petroleum gas*. Petroleum gas is a mixture of ethane, propane and butane. *But the main constituent of petroleum gas is butane* (C_4H_{10}).

Ethane, propane and butane are gases under normal conditions. Butane can be liquefied by applying pressure. *The petroleum gas which is liquefied under pressure is called Liquefied Petroleum Gas (LPG).*

The liquefied petroleum gas (LPG) consists mainly of *butane,* (with a very small amount of ethane and propane). Since the liquefied petroleum gas can be kept in the liquid from only under pressure hence LPG is stored and supplied in gas cylinders. *The gas cylinder we get from the gas agency (Indane, Burshane etc), contains liquefied petroleum gas (LPG).*

LPG is used

(i) As a domestic and industrial fuel

(ii) As a source of low molecular mass petrochemical.

COMPRESSED NATURAL GAS (CNG)

CNG, (Compressed Natural Gas) is highly compressed natural gas. Natural gas occurs deep inside the earth either alone or covering the crude oil in an oil field. Natural gas is a colourless, odourless mixture of methane (the main constituent), ethane, propane and butane.

Compressed natural gas is an excellent fuel. During burning it neither produces smoke nor any air pollutant. That is why, it is recommended to be used in automobiles in big cities.

The Octane Rating of Natural Gas is 130

A typical composition of natural gas is

Methane	Ethane	Propane	Butane	CO_2, N_2, H_2S
60–90%	5–9%	3–18%	2–14%	< 1% each

TEST ASSIGNMENT

Origin of coal and petroleum. Aromatic compounds from coal,. Fractional; distillation of coal tar. Hydrocarbons from petroleum. Cracking of petroleum, Reforming, Flash point of a hydrocarbon, Knocking of fuels, Octane number, Cetane number, Gasoline additives, Petrochemical, LPG, CNG.

VERY SHORT AND SHORT ANSWER QUESTIONS

1. Why is coal once again gaining important ?
2. Where does petroleum occur ?

GENERAL CHEMICAL PROPERTIES OF THE ALKANES

1. *Halogenation :* Chlorination may be brought about by light, hear or catalysts, and the extend of chlorination depends largely on the amount of chlorine used. A mixture of all possible isomeric monochlorides is obtained, but the isomers are formed

in unequal amounts, due to the difference of the reactivity of primary, secondary and tertiary hydrogen atoms. Markownikoff (1875) found experimentally that the order of ease of substitution is tertiary hydrogen > secondary > primary, *e.g.*, chlorination of isobutane at 300°C gives a mixture of two isomeric monochlorides.

$$CnH_{2n+2} \xrightarrow[-HX]{+X_2} C_nH_{2n-1}X \xrightarrow[-HX]{X_2} C_2H_{2n}X_2$$

$$\xrightarrow[-HX]{X_2} \text{... and so on.}$$

$$\underset{\text{(I)}}{CH_3\overset{\overset{CH_3}{|}}{C}HCH_2Cl} \quad \text{and} \quad \underset{\text{(II)}}{CH_3\overset{\overset{CH_3}{|}}{C}ClCH_3}$$

Isobutane has on tertiary and nine primary hydrogen atoms. One the basis of equal rates of substitution, the statistical results would be 90 per cent of (1) and 10 per cent of (II). Actually,, the results obtained were 67 percent of (1) and 33 percent of (II). Thus tertiary hydrogen is replaced about 4.5 times as fast as primary hydrogen.

Bromination is similar to chlorination, but not so vigorous. Iodination] is reversible, but it may be carried out in the presence of an oxidising agent, such HIO_3. HNO_3. HgO, etc., which destroys the hydrogen iodide as it is formed and so derives the reaction to the right, *e.g.*,

$$CH_4 + I_2 \; f \; CH_3I + HI$$

$$5HI + HIO_3 \rightarrow 3I_2 + 3H_2O.$$

Iodides are more conveniently prepared by treating the chloro- or bromo-derivative with sodium, iodide in methanol or acetone solutions. *e.g.*,

$$RCl + Nal \xrightarrow{\text{acetone}} Rl + NaCl$$

This reaction is possible because sodium iodide is soluble in methanol or acetone, whereas sodium chloride and sodium bromide are not. This reaction is known as the Finkelstein or Conant-*Finkelstein reaction (1910. 1924).*

Direct fluorination is usually explosive, special conditions are necessary for the preparation of the fluorine derivatives of the alkanes.

2. *Nitration :* Under certain conditions, alkanes react with nitric acid, a hydrogen atom being replaced by a nitro-group, NO_2.

This process is known as nitration. Nitration of the alkanes may be carried out in the vapour phase between 150° and 450°C, whereupon a complex mixture of mononitroalkanes is obtained. The mixture consists of all the possible mononitroderivatives and the nitro-compounds formed by every possibility of chain fission of the alkane, *e.g.*, propane gives a mixture of 1-nitropropane 2-nitropropane, nitroethane and nitromethane.

$$CH_3CH_2CH_3 \xrightarrow[400°C]{HNO_3} CH_3CH_2CH_2NO_2 + CH_3\underset{}{\overset{NO_2}{\overset{|}{C}}}HCH_3 + C_2H_5NO_2 + CH_3NO_2$$

As in the case of halogenation, the various hydrogen atoms in propane are not replaced with equal ease.

3. *Sulphonation :* Is the process of replacing a hydrogen atom by a sulphonic acid group. SO_3H. Sulphonation of a normal alkane from hexane onwards may be carried out by treating the alkane with oleum (fuming sulphuric acid) The ease of replacement of hydrogen atoms is. Tertiary very mush greater than secondary, and secondary greater than primary replacement of a primary hydrogen atom in sulphonation is very slow indeed. Isobutane, which contains a tertiary hydrogen atom, is readily sulphonated to give t-butylsulphonic acid.

$$(CH_3)_3CH + H_2SO_4 \longrightarrow (CH_3)_3CSO_3H + H_2SO_4$$

Sulphuryl chloride, in the presence of light and a catalyst, converts hydrocarbons into sulphonyl chlorides.

I has been shown that in concentrated suphuric acid, hydrocarbons containing a tertiary hydrogen atom undergo hydrogen exchange (Ingold *et al.*, 1936). The mechanism is believed to occur via a carbonium ion.

$$R_3CH + 2H_2SO_4 \longrightarrow R_3C^+ + HSO4^- + SO_2 + 2H_2O$$

$$R_3C+ + R_3CH \longrightarrow R_3CH + R_3C+.\ etc.$$

This reaction is of particular interest since optically active hydrocarbons have been racemised in sulphuric acid *e.g.*,. Burwell *et al.* (194*8) have shown that optically active 3-methyllhepatane is racemised in sulphuric acid.

4. *Oxidation :* All alkanes readily burn in excess of air or oxygen to from carbon dioxide and water. Controlled oxidation, on the other hand, under various conditions, leads to different products. Intensive oxidation gives a mixture of acids consisting of the

complete range of C1 to Cn carbon atoms. Less extensive oxidation gives a mixture of products in which no chain fission has occurred. Under moderate conditions, mixed ketones are the major products, and oxidation in the presence of boric acid produces a mixture of secondary alcohols. The oxidation of alkanes in the vapour state occurs via free radicals, *e.g.*, alkyl (R), alkylperoxy (ROO) and alkoxy (RO).

Oxidising reagents such as potassium permanganate readily oxidise a tertiary hydrogen atom to a hydroxyl group, *e.g.*, isobutane is oxidised to t-butanol.

$$(CH_3)_3CH + [O] \xrightarrow{KMnO_4} (CH_3)_3COH$$

5. *Isomerisation :* Of n-alkanes into branched-chain alkanes in which the side-chain is a methyl group, may be brought about by heating the n-alkane with aluminum chloride at 300°C, *e.g.*, n-butane isomerises into isobutane.

According to Pines *et al.* (1946), this isomerisation does not occur unless a trace of water is pressed (to from HCl) from the $AlCl_3$) together with a trace of alkyl halide or an alkane. The isomerisation is believed to be an ionic chain reaction. The alkene 'impurity' is converted into a carbonium ion (by the $AlCl_3$ and HCl), and this initiated the chain reaction, isomerisation then occurring by a 1.2 shift involving a hydride ion in the second step.

$$RCH{=}CH_2 + H^+ \rightleftharpoons R\dot{C}HCH_3$$

$$R\dot{C}HCH_3 + CH_3CH_2CH_2CH_3 \longrightarrow RCH_2CH_3 + \underset{CH_2\dot{C}HCH_3}{\overset{CH_3}{|}} \rightleftharpoons \underset{\dot{C}H_2C(CH_3)_2}{\overset{H}{|}} \rightleftharpoons (CH_3)_3\dot{C}$$

$$(CH_3)_3C^+ + CH_3CH_2CH_2CH_3 \rightleftharpoons (CH_3)_3CH + CH_3CH_2\dot{C}HCH_3$$

The formation of an equilibrium mixture of n- and isobutane from n-butane with the iso-compound predominating may be explained by the fact that the t -carbonium ion is more stable than the s-carbonium ion.

6. The thermal decomposition of the alkanes (see Cracking, below).

Some Individual Alkanes

Methane : This occurs in 'natural gas' (q.v) and the gases from oil wells. Methane is the principle product of organic decay in swamps and marshes, the gas being set free by the action of bacteria, this method of formation in nature has given rise to the name 'marsh gas' for methane. Sewage sludge which has been fermented by bacteria yields a gas containing about 70 per cent methane, and this is used as a liquid fuel. Methane also forms about 40 per cent by volume of coal gas..

Methane may be synthesised by striking an electric are between carbon electrodes in an atmosphere of hydrogen, or by heating a mixture of carbon and reduced nickel; at 475°C in the presence of hydrogen.

Sabatier and Senderns (1897) synthesised methane by passing a mixture of hydrogen and carbon monoxide or dioxide over finely divided nickel heated at about 300°C. *e.g.*,

$$CO + 3H_2 \longrightarrow CH_4 + H_2O.$$

Methane may conveniently be prepared by heating a mixture of anhydrous sodium acetate and soda-lime, or by reduction of methyl iodide with dissolving metals or with lithium aluminum hydride or lithium hydride. Other methods of preparation are by the action cf water on aluminum carbide or methylmagnesium iodide (a Grignard reagent).

$$Al_4C_3 + 12H_2O \longrightarrow 3CH_4 + 4\ Al(OH)_3$$

$$CH_3MgI + H_2O \longrightarrow CH_4 + MgI(OH)$$

Methane is obtained in vast quantities from natural gas, gas from the oil-wells, and from cracked petroleum (q. r.).

Properties of methane. Methane is a colourless, odourless, non-poisonous gas., its b.p. is – 164* C760 mm. and mp. – 184°C. It is somewhat soluble in water, 100 ml of water dissolving about 5 ml of methane at 20°C, but is quite soluble in ethanol and ether. It burns with a non-luminous flame in air or oxygen, forming carbon dioxide and water.

$$CH_4 + 2O_2 \longrightarrow CO_2 + 2H_2O$$

It explodes violently when mixed with air (or oxygen) and ignited, and this is believed to be the cause of explosions in coal-mines, where methane is known as free-damp. Methane may be catalytically oxidised to methanol and formaldehyde.

Uses of methane. Methane, by heating to 1000°C or by incomplete combustion, produces carbon in a very finely divided state. Carbon prepared this way is known as carbon black, and

$$C + 2H_2O \xleftarrow{O_2} CH_4 \xrightarrow{1000°C} C + 2H_2$$

is used to make paints and printers' ink, it is also used in the rubber industry for motor tyres, etc.

Methane is used as a source of synthesis gas. A mixture of methane and steam is passed over heated nickel (800°C) supported on alumina.

$$CH_4 + H_2O \longrightarrow CO + 3H_2, \; CH_4 + 2H_2O \longrightarrow CO_2 + 4H_2.$$

The reaction to give carbon dioxide occurs to a much less extent than that to give carbon molecule, and the carbon dioxide is removed from the mixture by washing with water under pressure.

Methane is also used as a fuel and as the starting point of many chemicals 2. *Ethane.* C_2H_6, occurs with methane in natural gas and the gases from the oil-wells. It may be prepared by the action of dissolving metals on ethyl iodide, by a Wurtz reaction with Mel, by the electrolysis of sodium acetater, or by the action of water on Et Mgl. The best way of preparing pure ethane in large quantity is by the catalytic reduction of ethylene.

Properties of ethane : Ethane is a colourless gas, b.p. – 89) C, sparingly soluble in water but readily soluble in ethanol. It burns in air or oxygen to form carbons dioxide and water. It reacts with halogens in a similar manner to methane to form substitution products, but a much greater number of products are possible due, firstly to the presence of six hydrogen atoms in ethane compared with four in methane, and secondly, to the fact that isomerism is passible in the substitution products of ethane, and not in those of methane. Thus for example, two dichloropethanes are possible. CH_3CHCl_2 and $ClCH_2CH_2Cl$. Ethane is mainly used as a fuel.

Propane : C_3H_5. is a constituent of natural gas and gas from the oil-wells. It may be prepared by methods used for ethane, using the appropriate starting materials. It is a colourless gas. b.p. – 44.5°C, and is used as a fuel.

Butenes : C_4H_{10} Both n- and isobutane occur in natural gas and petroleum gas, and they may be separated by tractional distillation under pressure. The butenes are used as a liquid fuel and for the preparation of various compounds.

n-Butane. b.p. – 0.5°C, may be prepared by the Wurtz reaction.

$$2CH_3CH_2I + 2Na \longrightarrow CH_3CH_2CH_2CH_3 + 2NaI$$

Solbutane, b.p., – 10.2°C, may be prepared by treating t-butyl iodide with a zinc-copper couple in ethanol.

$$(CH_3)_3 CI \xrightarrow{\text{a H*}} (CH_3)_3 CH + HI$$

Pentanes, C_5H_{12} : All three pentanes occur in natural gas and petroleum gas and are used for preparing chloropentanes and the pentanols.

n-Pentane. $CH_3CH_2CH_2CH_2CH_3$ b.p., 36° C. isopentane, $(CH_3)_2CHCH_2CH_3$. b.p... 28° C. neopentane, $(CH_3)_4C$bp. 94°C

Petroleum and Naturals Gas

Crude petroleum (mineral oil) is the term usually applied to the gases accruing naturally in the oil fields, the liquid from the wells, and the solids which are dissolved in, or have been separated from, the liquid. The composition of crude petroleum varies with the locality of occurrence, but all contain alkanes (straight and branched-chain from about C_1 to C_{40}), cycloalkanes or napthens, and aromatic hydrocarbons. The low-boiling fractions of almost all petroleums are composed of alkanes, it is the composition of the higher-boiling fractions which differs according to the sources of the petroleum. In addition to hydrocarbons, there are also present compounds containing oxygen, nitrogen, sulphur and metallic constituents.

Distribution and general composition of crude petroleum. If the residue of petroleum, after removal of volatile compounds, contains a large amount of alkanes or wax, the petroleum is classified as paraffinic or paraffin base oil. If naphthenes predominate, the petroleum is classified as asphaltic or asphalt base oil. The crudes from the wells in Pennsylvania. Iran. Irak and Rumaina are paraffinic, those from Baku and Venezuela are asphaltic, and those from Oklahoma. Texas and Mexico are intermediate in composition, and may be classified as paraffinic and asphaltic.

Distillation of petroleum : The crude oil is nearly always associated with water and sand. Hence the crude petroleum discharged from the top of the well contains water and sand in suspension. The mixture is passed, under pressure, into cylindrical tanks, and the gas, oil and solids are drawn off separately.

Except for the boiling hydrocarbons, no attempt is made to separate the individual hydrocarbons. The crude oil is fractionated by continuous distillation into four main fractions. Petrol (gasoline), kerosene (kerosine, paraffin oil), gas oil (heavy oil) and lubricating oil. The residue may be fractionated by means of vacuum-distillation to give light, medium and heavy lubricating oils, paraffin wax, and asphaltic bitumen. Each of the four main fractions may be further split up by batch distillation into fractions of narrow boiling range. Recently,, it has been possible to isolate individuals by 'super fractionation'. The final number of fractions taken depends on the purpose in view.

Table 1.10

Name	*Bp°C*	*Approximate composition*	*Uses*
Light petrol	20– 100	C_5H_{12}-C_7H_{16}	Solvent
Benzine	70 – 90	C_6–C_7	Dry cleaning
Ligroin	80 – 120	C_6–C_8	Solvent
Petroleum (gasoline)	70 – 200	C_6–C_{11}	Motor fuel
Kerosene (paraffin oil)	200 – 300	C_{12}–C_{16}	Lighting
Gas oil (heavy oil)	above 300	C_{13}-C_{18}	Fuel oil
Lubricating oil (mineral oil)	"	C_{16}-C_{20}	Lubricants
Greases, vaseline, petrolatum	"	C_{18}-C_{22}	Pharmaceutical preparations
Paraffin wax (hard wax)	"	C_{20}-C_{30}	Candles, waxed paper. etc.
Residue (asphaltic bitumen)	"	C_{30}-C_{40}	Asphalt tar, petroleum coke

Refining of the various fractions : It appears that refining was originally introduced to remove the bad colour and objectionable odour of petrol. Today it is realised that it is more important to remove sulphur compounds which lower the response of petrol to added tetraethyl-lead.

An internal-combustion engine, *i.e.*, one which burns fuel within the working cylinder, is ore efficient the higher the compression ratio. Petrol engines use 'spark ignition', and as the compression ratio increases, a point is reached when 'knocking' is observed, *i.e.*, after passage of the firing spark, instead of all the fuel gas burning smoothly, the end portion burns with explosive violence, given rise to a metallic rattle. The phenomenon of knocking is still not fully understood, but it has been found that, among other factors, the tendency to knock depends on the nature of the petrol. n-Alkanes tend to produce knocking far more than branched-chain alkanes. Edgar (1927) introduced 2.24 trimethylpentane (incorrectly known as iso-octane), which has higher anti-knock properties, and n-heptane, which has lower anti-knock properties than any commercial petrol, as standards for rating fuels Iso-octane is arbitrarily given the value of 100 and -heptane, 0 and the octane number of any fuel is the per cent of iso-octane' in a mixture of this compound and n-heptane which will knock under the same conditions as the fuel being tested.

Alkanes and aromatic compounds have high octane numbers. Tetraethyl-lead also raises the octane number of a given petrol. but if sulphur compounds are present, the response to this 'dope' is lowered. Hence it is very important to remove sulphur compounds from petrol. The method of refining depends on the particular fraction concerned, and it is not practicable to refine before distillation (of the petroleum).

Cracking

The thermal decomposition of organic compounds is known as pyrolysis, pyrolysis, when applied to alkanes, is known as creaking.

When heated to about 500-600°C, alkanes are decomposed into smaller molecules and the products obtained from a given alkane depend on.

(i) the structure of the alkane.

(ii) the pressure under which cracking is carried out, and

(iii) the presence or absence of catalysts such as silica-alumina, silica-alumina-thoria, silica-alumina-zirconia.

The mechanism of cracking is still observe. Many theories have been suggested, and one that is highly favoured is a free-radical mechanism, evidence for which has been obtained from the observation that at cracking temperatures many hydrocarbon produces alkyl free radicals.

When petroleum is cracked, of all the compounds produced., the most important are those containing up to four carbon atoms, methane, ethane ethylene, propane, propene, butane, butene and isobutene. All of these have found wide application as the materials for the preparation of a large number of chemicals.

By using suitable catalysts, alkanes containing six or more carbons atoms may be catalytically cyclised, *e.g.*, n-hexane. Under pressure, passed over chromic oxide carried on an alumina support and heated at 480-550°C, gives benzene.

$$C_6H_{14} \longrightarrow C_6H_6 + 4H_2$$

Therefore are two main types of cracking. (i) liquid phase, and (ii) vapour phase.

(i) *Liquid phase cracking :* Heavy oil (from the petroleum distillation) is cracked by heating at a suitable temperature (475-530°C) and under pressure (100-1000 1b/sq. in), by means of which the cracked metrical is maintained in the liquid condition. The heavy oil is converted into gasoline to the extent of 60-65 per cent of the oil (by volume), and has an octane number of 65-70. If attempts are made to increase the yield of gasoline, the octane number decreases.

(ii) *In vapour phase cracking :* The cracking temperature is 600°C and the pressure is 3-5-10.5kg/cm^2. The cracking stock any be gasoline, kerosene, gas oils,, but not the heavy oils, since these cannot be completely vaporised under the above conditions.

Reforming is the process whereby straight-run gasoline is cracked in order to raise the octane number. The gasoline is heated to about 600°C and under pressure of 28-52.5kg/cm^2 and the yield varies from 60 to 90 percent, the greater the yield , the lower is the octane number. Catalysts – the oxides of silicon and aluminium, plus small amounts of other oxides such as magnesia, zirconia, etc. – are usually employed in the reforming process. The chief chemical changes that occur are isomerisation, dehydration and cyclisation.

Catalytic cracking is also increasing in use, since it has been found that catalytically cracked gasoline contains few alkanes that readily polymerise. Gum formation in cracked gasolines prevented by the addition of inhibitors, which are mainly phenols or aromatic amines, *e.g.*, catechol, p-benzyl-aminophenol, naphthylamine.

Treatment of natural gas : When natural gas does not contain hydrocarbons above ethane, it is said to be 'lean' or 'dry 0146, when it contains the higher hydrocarbons (up to hexane) it is said to be 'rich' or 'wet'. The alkanes may be separated by fractional distillation under increased pressure, thereby giving methane, ethane, propane, n-and isobutanes, n, iso- and neopentanes, and hexane. These gases are used for various purpose. On the other hand, natural gas itself may be used in the manufacture of various compounds. Oxidation of natural gas under carefully controlled conditions produces a complex mixture of compounds, among which are formaldehyde, acetaldehyde, acetic acid acetone, methanol, ethanol, propanols and butanols. These are separated by distillation, solvent extraction, etc.

We has is also used as a source of gasoline. The vapours of the liquid hydrocarbons (penances and hexane) in the wet gas are removed by various methods, *e.g.*, compression and cooling. The liquid product obtained from wet gas is known as 'natural' of 'casinghead' gas, and is 'wild' because of the dissolved gases, in it. These gasses may be removed by distillation under pressure, and the resulting liquid is known as 'stabilised natural gasoline', this has very high anti-knock properties.

$$xCO + yH_2 \longrightarrow \text{mixture of hydrocarbons} + \text{water}$$

Various metals and oxides have been used as catalysts. One of the best is, cobalt (100 parts), thoria (5 parts) magnesia (8 parts) and kieselguhr (200 parts). When synthesis gas is passed over the catalyst at moderate pressure (9-11 atm), more of the hugh-boiling fraction is obtained than when the process is carried out at atmospheric pressure. The liquid products are fractionally distilled, and refined in the same way as are the petroleum fractions, furthermore the higher-boiling fractions are cracked.

Gasoline from the synthine process costs more than that from petroleum. The Fisher-Tropsch oils appear to be more valuable as chemical raw materials than as fuels, and have been used for the production of higher alkanes, monocarboxylic acids, detergents and in the oxo-process.

SOLVED EXAMPLES

Example 1:

Draw the Sawhorse and Newman projection of butane using the C – 2 to C – 3 bonds as references in

(i) eclipsed form *(ii) staggered form*

Solution:

CH_3 CH_3 CH_3 CH_3

H H H H H H H H H H

Sawhorse Newman

(i) Eclipsed form of butane

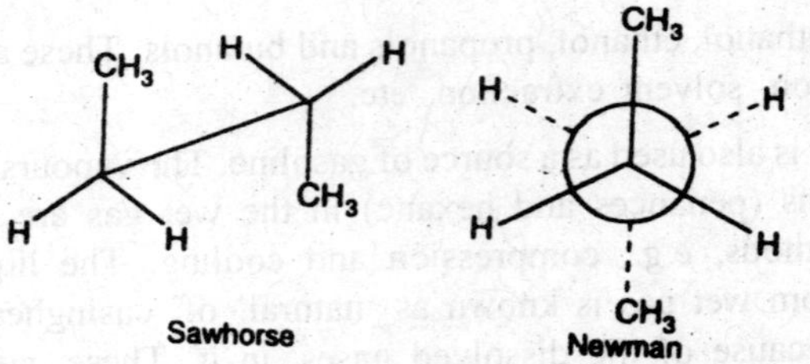

Sawhorse Newman

(ii) Staggered form of butane

Fig. 1.26

Example 2:

Why is cyclopropane very reactive when compared to cyclohexane?

Solution:

The $C\hat{C}C$ angle in cyclopropane is 60° and in cyclohexane 120°. So the angle strain in the two molecules is,

$$\text{Angle strain in cyclopropane} = \frac{1}{2}(109° \, 2' - 60°) = 24°44'$$

$$\text{Angle strain in cyclohexane} = \frac{1}{2}(109° \, 28' - 120°) = -5°16'$$

The deviation from the normal tetrahedral angle is more in cyclopropane than in cyclohexane. Thus, the cyclopropane molecule is highly strained molecule as compared to cyclohexane molecule. As a result, cyclopropane is more reactive than cyclohexane.

Example 3:

Explain the term 'knocking'. What is the relationship between the structure of a hydrocarbon an knocking?

Solution:

The metallic sound produced by the explosive combustion of gasoline-air mixture inside the cylinder of an internal combustion engine is called knocking.

The knocking tendency of hydrocarbons decreases as follows.

Straight chain alkanes > Branched chain alkanes > Alkanes > Arenes

Example 4:

Describe different methods to improve the quality of a fuel used in a gasoline engine.

Solution:

The quality of gasoline can be improved as follows

(i) By adding tetraethyl lead to the gasoline.

(ii) By adding alkanes, branched-chain alkanes or aromatic hydrocarbons to gasoline. These compounds are generally prepared by catalytic cracking and catalytic reforming.

EXERCISES

1. Draw Newman projection formula for the eclipsed and staggered forms of. (a) ethanol. (b) 1-bromo-2-chloroethane. (c) propane, (d) 1.2-dibromopropane.
2. An alkane has M.W. of 72, and monochlorination produces one product only. What is the compound ?
3. ΔG for the formation of n-butane is – 15.69 kJ mol^{-1}, and for isobutane is – 17.99kJ mol^{-1}, and for isobutane is – 17.99 kJ mol^{-1}. Starting with 1 mole of n-butane, calculate the moles of n- and isobutane in the mixture after heating in the presence of a catalyst) until equilibrium is obtained.
4. How many signals can be expected in the NMR spectrum of $MeCHBrCH_2Br$?. Give reasons for your answer.

   N. B. The reader should test himself on the general methods of preparation and general properties of any homologous series.
5. $Me_2CO \longrightarrow Me_2CH_2$:(ii) $PrBr \longrightarrow PrH$: (iii) $Me_3CH \longrightarrow Me_3COH$, (iv) $MeCH{=}CHMe \longrightarrow MeCH_2.CH_2Me$ (v) MeI + EtI $\xrightarrow{Na}$? (vi) EtMgI $\xrightarrow{H_2O}$? (vii) $EtCO_2^- + e \longrightarrow$? (viii) $EtCO_2^-Na^+ \xrightarrow{heat}$? (ix) $EtCH{=}CH_2 \xrightarrow{B_2H_4}$? $\xrightarrow[NaOH]{AgNO_3}$?
6.

m/e	15	27	29	39	41	42	43	44	56	57	58	72
A(%)	7	41	46	21	68	87	100	33	17	55	24	6
B (%)	–	34	25	14	41	59	100	34	–	13	6	9
C (%)	5.3	15	–	13	42	–	–	–	–	100	44	–

2

ALKENES AND ALKYNES

ALKENES OR OLEFINS

The alkenes are the unsaturated hydrocarbons that contain one double bond. They have the general formula C_nH_{2n}, and the double bond is known as the 'olefinic bond' or 'ethylenic bond'.

The name olefin arose form the fact that ethylene was called 'olefiant gas' (oil-forming gas), since it formed oily liquids when treated with chlorine or bromine. The original name given to this homologous series was olefine ; but it was later decided to reserve the suffix *-ine* for *basic* substances only. Since the name olefine had gained wide usage, it was decided to compromise and call the series the olefins.

One method of nomenclature is to name the olefin from the corresponding alkane by changing the suffix *-ane* of the latter into *-ylene, e.g.*, |(methylene), ethylene, propylene, etc.

Isomers differing only in the position of the double bond are prefixed by Greek letters or numbers which indicate the position of the double bond. The lowest number is usually given to the double bond (see below), and the number (or Greek letter) indicates the first of the two carbon atoms that are jointed together by the double bond.

Another method of nomenclature is to consider ethylene as the parent substance and the higher members as derivatives of ethylene. If the compound is a monosubstituted derivative of ethylene, then no difficulty is encountered in naming it; if the compound is a disubstituted derivative of ethylene isomerism is possible, since the alkyl groups can be attached to the same or different carbon atoms. When the groups are attached to the same carbon atom the olefin is named as the *symmetrical* or *unsymmetrical* compound (abbreviated to *as-* or *unsym*); when attached to different carbon atoms the olefin is named as the *symmetrical* (*sym*- or *s*-) compound.

According to the I.U.P.A.C. system of nomenclature, the class suffix of the olefins is *-ene*, and to the series becomes the alkene series. The longest carbon chain containing the double bond is chosen as the parent alkene, the name of which is obtained by changing the suffix *-ane* of the corresponding alkane into *-end*. The positions of the double bond and side-chains are indicated by numbers, the lowest number possible being given to the double bond, and this is placed before the suffix. The I.U.P.A.C. system is not the one generally used.

NOMENCLATURE

Various member of alkene family are :

C_1H_2 → does not exist

C_2H_4 → Ethene (ethylene) $H_2C = CH_2$

C_3H_6 → Propene (Propylene) $CH_3—CH = CH_2$

C_4H_8 → Butene (Butylene) $CH_3—CH_2—CH = CH_2$

In general, alkenes can be represented by the formula:

$$\begin{matrix} R & & R \\ & C=C & \\ R & & R \end{matrix}$$

$CH_3CH_2CH = CH_2$ but-1-ene (1-butylene, a-butylene, ethyletylene)

$\overset{3}{CH_3}—\underset{\displaystyle CH_3}{\overset{2}{C}}=\overset{1}{CH_2}$ 2-methylprop-1-ene (isobutene, isobutlene, as-dimethylethylene)

$\overset{1}{CH_3}\overset{2}{CH}=\overset{3}{C}(CH_3)\overset{4}{CH_2}\overset{5}{CH_3}$ 3-methylpent-2-ene (3-methyl-2-pentylene)

$CH_3CH_2\underset{\displaystyle CH_2}{\overset{2}{C}} — \underset{\displaystyle CH_3}{\overset{3}{C}}\overset{4}{H}CH_3$ 2-ethyl-3-methylbut-1-ene

When there are several chains of equal length containing the double bond, then the same principles apply as for the alkanes (see last example).

The names of univalent groups derived from alkenes have the endingenyl, and the carbon atom with the free valency is numbered 1. There are three exceptions.

$\overset{3}{C}H_3\overset{2}{C}H = \overset{1}{C}H—$ prop-1-enyl

$\overset{5}{C}H_3\overset{4}{C}H(CH_3)\overset{3}{C}H = \overset{2}{C}H\overset{1}{C}H_2—$ 4-methylpent-2-enyl

$CH_2 = CH—$ vinyl (for ethenyl)

$CH_2 = CHCH_2—$ allyl (for propen-2-yl)

$CH_2 = C(CH_3)—$ isopropenyl (for 1-methylvinyl)

A double or triple bond is regarded as a functional group. When there are several functional groups in the molecule, then the lowest number possible is given, in order of preference :

(*i*) To the *principal* functional group of the compound :

(*ii*) To the double or triple bond; and

(*iii*) To atoms groups designated by prefixes.

GENERAL METHODS OF PREPARATION OF THE ALKENES

By passing the vapour of an alcohol over Alumina (Al_2O_3) at 623K (350°C)

$$R—CH_2—CH_2CH \xrightarrow[-H_2O]{Al_2O_3\ 623\,K} \underset{\text{alkene}}{R—CH = CH_2}$$

(*a*) By the action of concentrated sulphuric acid, at 160-170 C, on primary alcohols. The acids act as a dehydrating agent, *e.g.*, ethylene from ethanol :

$$C_2H_5OH^+ \xrightarrow{H_2SO_4} C_2H_5OH_2^+ \longrightarrow H_2O + C_2H_5^+ \longrightarrow C_2H_4 +$$

H^+ (*f.–g.*)

Dehydration of secondary and territory alcohols is best carried out using *dilute* sulphuric acid since the alkenes produced from these alcohols (particularly tertiary alcohols) tend to polymerise under the influence of the concentrated acid. The yields of alkenes from secondary and tertiary alcohols are very good. Also, the ease of dehydration of alcohols is t > s > primary.

Acid-catalysed dehydration of primary alcohols normally gives 1-alkenes, but s- and t-alcohol give mixtures of alkenes which arise from removal of a proton from either adjacent carbon atom (to the COH group) and also from rearrangements undergone by the various carbonium is intermediates. Rearrangements, however, may be avoided by dehydrating the alcohol over heated alumina in the presence of a little pyridine.

An advantage of cyclic eliminations is that no carbon skeleton rearrangement occurs (*cf.* dehydration of alcohols, method 1). Also, in the pyrolysis of esters and xanthates of s- and t-alcohols, mixtures of alkenes are produced, but the terminal alkene is favoured over the non-terminal alkene.

(a) *Pyrolysis of esters* : The esters are usually acetates:

$$R_2C(H)-CH_2-O-C(=O)Me \xrightarrow{500\ C} [\text{cyclic transition state}] \longrightarrow R_2C{=}CH_2 + HO-C(=O)Me \quad \text{(v.g.)}$$

In many cases, pyrolysis of the alcohol with acetic anhydride is simpler than starting with the preformed ester (Aubrey *et al.*, 1965).

(b) *Pyrolysis of xanthates* : This is known as the *Tschugaev* (*Chugaev*) *reaction* (1899).

$$R_2CHCH_2OH \xrightarrow[NaOH]{CS_2} R_2CH-CH_2-O-C(=S)S^-Na^+ \xrightarrow{MeI} R_2CH-CH_2-O-C(=S)SMe \xrightarrow{200°\ C}$$

$$[\text{cyclic transition state}] \longrightarrow R_2C{=}CH_2 + [HS-C(=O)SMe] \longrightarrow MeSH + COS \quad \text{(v.g.)}$$

The yield of alkene is in increased by heating the xanthate with, *e.g.*, chloroacetic acid (—CS^- Na^+—$CSCH_2CO_2H$), and then refluxing this is feebly acid solution. Alternatively, heating the xanthate derivative in the presence of a Lewis acid catalyst, *e.g.*, BF_3, shortens the reaction time and the yield is increased.

(c) *Cope reaction (1949)* : This the reaction in which alkenes are formed when amine oxides (p. 378) are heated :

$$RCH(H)-CH_2-\overset{+}{N}Me_2-O^- \xrightarrow{150\ C} [\text{cyclic transition state}, O^{\delta-}, N^{\delta+}] \longrightarrow RCH{=}CH_2 + Me_2NOH$$

The reaction may also be carried out in dimethyl sulphoxide or tetrahydrofuran at room temperature (Cram *et al.*, 1962).

(b) By the action of *ethanolic* potassium hydroxide on alkyl halides, *e.g.*, propene from propyl bromide:

$$CH_3CH_2CH_2Br + KOH \xrightarrow{\text{ethanol}} CH_3CH = CH_2 + KBr + H_2O$$

This is not a very important method for the preparation of the lower alkenes, since these may be prepared directly from the corresponding alcohols, which are readily accessible. The reaction however, is very important, since by means of it a double bond can be introduced into an organic compound. The yield of a alkene depends on the nature of the alkyl halide used; it is fair with primary, and very good for secondary and tertiary alkyl halides. Ethylene cannot be prepared by this method from ethylhalides for the mechanism of *dehydrogenation, i.e.*, removal of halogen acid.

Dehalogenation (*i.e.*, removal of halogen) of 1,1-dihalogen derivatives of the alkanes by means of zinc dust and methanol produces alkenes, *e.g.*, propene from propylidene dibromide :

$$CH_3CH_2CHBr_2 + Zn \longrightarrow ZnBr_2 + [CH_3CH_2\ddot{C}H] \longrightarrow CH_3CH{=}CH_2$$

If sodium is used instead of zinc, and the reaction is carried out preferably in ether solution, comparatively little propene is formed the main product being hex 3-ene.

$$2CH_3CH_2CHBr_2 + 4Na \longrightarrow CH_3CH_2CH = CHCH_2CH_3 + 4NaBr$$

This reaction is really an extensión of the Wurtz synthesis, and the important point to note is that the use of sodium tends to produce lengthening of the carbon, chain.

Zinc dust and methanol also dehalogenate 1,2-dihalogen derivatives of the alkanes, *e.g.*, propene from propene dibromide :

$$CH_3CHBrCH_2Br + Zn \longrightarrow CH_3CH = CH_2 + ZnBr_2$$

A possible mechanise is :

$$Zn \longrightarrow Zn^{2+} + 2e$$

$$2e + Br-\overset{|}{\underset{|}{C}}-\overset{|}{\underset{|}{C}}-Br \longrightarrow Br^- + -\overset{|}{\underset{|}{\ddot{C}}}-\overset{|}{\underset{|}{C}}-Br \longrightarrow \rangle C{=}C\langle + Br^-$$

Neither method is useful for preparing alkenes, since the necessary dihalogen compounds are not readily accessible. The second method, however, is useful for purifying alkenes or for 'protecting' a double bond (see, *e.g.*, allyl alcohol). Sodium iodide may be used instead of zinc dust, and according to Mulders *et al.* (1963), the mechanism is :

$$\mathrm{I^- + \rangle C(Br){-}C(Br)\langle \xrightleftharpoons{fast} Br^- + \rangle C{-}\overset{+}{Br}{-}C\langle\ (\text{bridged}) \xrightarrow{slow} \rangle C{=}C\langle + IBr}$$

I

Other means of removing the two bromine atoms to regenerate the alkene are chromous chloride (Allen *et al.*, 1960), chromous sulphate (Castro *et al.*, 1963), or sodium trimethoxyborohydride (King *et al.*, 1968).

Both *addition* (to the alkene) and *elimination* (from the dibromide) of the two bromine atoms are predominantly *trans, i.e., stereoselective,* and these observations may be explained in terms of the formation of a bridged intermediate. It also follows from this that the configuration of the 'protected' alkene remains unchanged.

(c) By heating a quaternary ammonium hydroxide, *e.g.*, ethylene from teraetylammonium hydroide (see p. 379) :

$$(C_2H_5)_4N + OH^- \longrightarrow C_2H_4 + (C_2H_5)_3N + H_2O$$

Boord *et al.* (1930-33) have prepared alkenes by conversion of an aldehyde into its chloro-ether, treating this with bromine followed by a Grignard reagent and finally treating the product with zinc and n-butanol.

$$R^1CH_2CHO \xrightarrow{C_2H_5OH} R^1CH_2CHClOC_2H_5 \xrightarrow{Br_2} R^1CHBrCHBrOC_2H_5$$

$$\xrightarrow{R^2MgHr} R^1CHBrCHR_2OC_2H_5 \xrightarrow{Zn} R^1CH = CHR^2$$

This method is very useful for preparing alkenes of definite structure, and an interesting point about it is the replacement of the α-chlorine atom by bromine when the α-chloro-ether undergoes bromination in the β-position. The greater reactivity of the α-bromine atom may e explained as follows :

$$\mathrm{RCHBr - CH(Br) - \ddot{O} - C_2H_5 \longrightarrow RCHBrCH{=}\overset{+}{O} - C_2H_5 + Br^-}$$

The Wittig reaction. This is also a means of preparing alkenes where the position of the double bond is definite.

A number of alkenes are prepared by the cracking of petroleum, *e.g.*, ethylene, propene, butenes, etc. (see also the individuals). For the production of the lower alkenes the most suitable starting material is gas oil, whereas the higher alkenes it is best to use paraffin wax or Fischer-Triopsch wax. The loser alkenes (C_2—C_5) are also prepared by the catalytic dehydrogenation of saturated hydrocarbons, the most satisfactory catalysts being those of the chromium oxide-alumina type.

General properties of the alkenes

(*a*) *Physical state* : The members containing two to four carbon atoms are gases; five to seventeen, liquids; eighteen onwards, solids at room temperature and they burn in air with a luminous smoky flame.

(*b*) *Density* : Alkenes are lighter than water.

(*c*) *Solubility* : Alkenes are insoluble in water and soluble in organic solvents, such as benzene other etc.

(*d*) *Boiling print* : The boiling points of alkenes show a gradual increase with an increase in the molecular mass (or chain length) this indicate that the inter molecular attractions become and stronger with the increase in the size of molecule.

Geometrical isomers (*cis* and *trans*), however, behave differently because of their different geometry.

Owing to the presence of a double bond, the alkenes undergo a large number of *addition* reactions, but under special conditions they also undergo substitution reactions.

The high reactivity of the olefinic bond is due to the presence of the two *n*-electrons, and when addition occurs, the trigonal arrangement in the alkene changes to the tetrahedral arrangement in the saturated compound produced.

Alkene addition reactions involve an *electrophile* in the *first* step. This is the rate-determining step, and the intermediate, a carbonium ion, then reacts rapidly with a *nucleophile.* The energy profile of a two-step reaction of this type. Also, the more stable the carbonium ion, the faster is the addition.

If E and Nu represent, respectively, the electrophilic and nucleophilic 'fragments' of the addendum E—Nu, the addition may be generalised as follows :

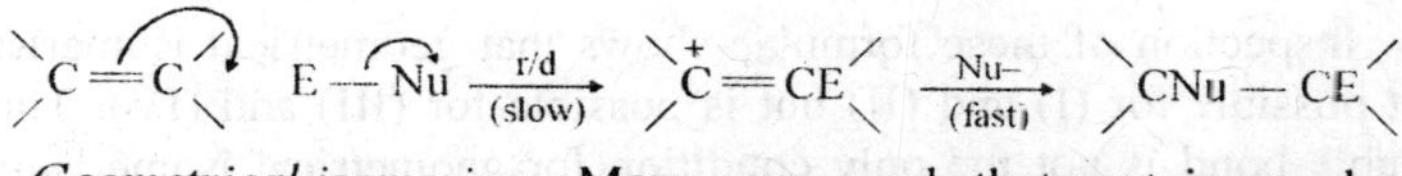

$$>C{=}C< \;+\; E{-}Nu \xrightarrow[\text{(slow)}]{\text{r/d}} >\overset{+}{C}{-}CE< \xrightarrow[\text{(fast)}]{Nu^-} >CNu{-}CE<$$

Geometrical isomerism : Many compounds that contain one double bond exist in two forms which differ in most of their physical and in many of their chemical properties, can't Hoff suggested that if we assume they is *no free rotation about a double bond,* two spatial arrangements are possible for the molecule (*ef.* rotation about a single bond.). If we assume that the two unsaturated carbon atoms are sp^3 hybridised, then they are joined by two 'banana' bonds. The two banana bonds are formed by the maximum overlap possible of the sp^3 orbitals (in pairs), and to decrease this overlap (in either bond), energy must be supplied to break the bond. Thus there is resistance to rotation about a double bond. On the other hand, if we assume sp^2 hybridisation for the two unsaturated carbon atoms, then in this case it is the maximum overlap of the two π-orbitals which causes resistance to the rotation about the double bond. Furthermore, either way of representing the molecule produces a *planar* spatial arrangement.

Isomers which have the same structure but differ in their spatial arrangement, *i.e.*, have different *configurations*, are said to be *stereoisomers*. This type os isomerism is called stereoisomerism, and is of two kinds: optical isomerism and geometrical isomerism (or *cis-trans* isomerism). In optical isomerism, all (or at least some) of the isomers are optically active. On the other hand, geometrical isomerism is characterised by the existence of different configurations for a given structure which may or may not be optically active, *i.e.*, optical activity is not a criterion for geometrical isomerism but is, *by definition*, a criterion for optical isomerism.

Several types of structures are capable of exhibiting geometrical isomerism, but we shall here discuss the one in which the molecule contains a double bond. Such compounds may be of various types, *e.g.*, $Ca = Cb_2$, $Ca_2 = Cbd$, $Cab = Cab$, $Cab = Cad$. These are drawn conventionally at (I)–(IV).

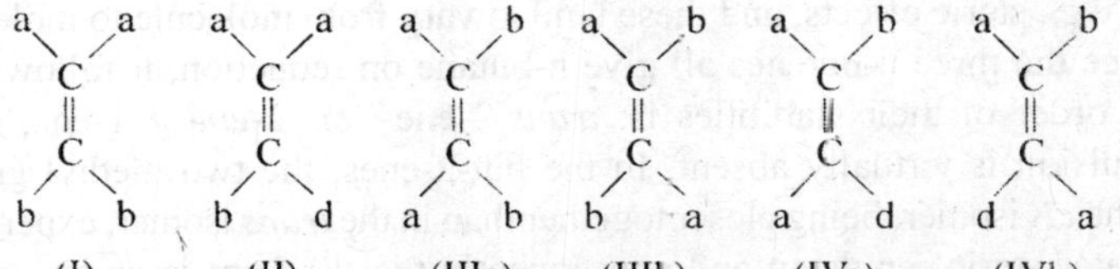

(I)	(II)	(IIIa)	(IIIb)	(IVa)	(IVb)
a a	a a	a b	a b	a b	a b
C	C	C	C	C	C
‖	‖	‖	‖	‖	‖
C	C	C	C	C	C
b b	b d	a b	b a	a d	d a

Inspection of these formulae shows that geometrical isomerism is not possible for (I) and (II) but is possible for (III) and (IV). Thus, a double bond is not the only condition for geometrical isomerism; the groups attached to the unsaturated carbon atoms must also be taken into consideration.

Geometrical isomers are named as *cis* and *trans* compounds, respectively, according as identical (or similar) atoms or groups are on the *same* side or *opposite sides, e.g.*, but-2-enes:

Me(H)C=C(H)Me	Me(H)C=C(Me)H
cis	*trans*

The two terms *conformation* and *configuration* often lead to confusion, since both are concerned with spatial arrangements. They may be differentiated as follows: change from one configuration to another involved breaking and making of bonds, whereas change from one conformation to another involved rotations (or twisting) about a single bond of one part of the molecule with respect to another part.

Stabilities of alkenes. One way of measuring the stability of an alkene is the determination of its heat of hydrogenation, *e.g.*, (ΔH in kJ mol^{-1})

$CH_2 = CH_2$	$MeCH = CH_2$	$MeCH_2CH = CH_2$
–137.2	–125.9	–126.8
$MeCH = CHMe$		$Me_2C = CH_2$
cis, –119.7; *trans,* –155.5		–118.8

Since the reaction is exothermic, the smaller ΔH is (numerically), the more stable is the alkene *relative* to its parent. Thus, it is only possible to compare the stabilities of different a alkenes which produce the *same* alkane on hydrogenation. This arises from the fact that the enthalpy of formation of alkanes is not a purely additive property; it also depends on, *e.g.*, steric effects, and these tend to vary from molecule to molecule, Since the three n-butenes all give n-butane on reduction, it follows that the order of their stabilities is: *trans* 2-ene> *cis* 2-*ene* > 1-ene, steric repulsion is virtually absent. In the but-2-enes, the two methyl groups in the *cis* isomer, being closer together than in the *trans* isomer, experience greater steric repulsion and consequently the *cis* form is under greater strain than the *trans*. Thus steric repulsion destabilises a molecule. On the other hand, hyperconjugation stabilises a molecule, and is small in but-1-ene but much larger in the but-2-enes. Since *trans*-but-2-ene is the most stable isomer, it followers that hyperconjugation has a greater stabilising effect steric repulsion a destabilising effect (in these three butenes).

Stabilities of a alkenes may also be compared by the determination of their heats of combustion (exothermic reaction), *e.g.*, (ΔH in kJ mol^{-1}):

$MeCH_2CH = CH_2$	$MeCH = CHMe$	$Me_2C = CH_2$
–2719	*cis*, –2712; *trans*, -2707	–2703

In this case all four butenes may be compared, since all give the *same products on combustion, viz.* $4CO_2+4H_2O$. The order of stabilities is thus: iso > *trans* 2-ene > *cis* 2-ene > 1-ene.

In general, the order of stability of alkenes is:

$R_2C = CR_2 > R_2C = CHR > R_2C = CH_2 \sim RCH = CHR$ (*trans* > *cis*) $> RCH = CH_2 > CH_2 > CH_2$

Reaction of alkenes

1. Alkenes are readily hydrogenated under pressure in the presence of a catalyst. Finely divided platinum and palladium are effective at room temperature; nickel on a support (Sabatier-Senderens reduction) requires a temperature between 200° and 300°C; Raney nickel is effective at room temperature and atmospheric pressure

$$CnH_2N+H_2 \xrightarrow{\mu\text{-}250^\circ C} \text{en } H_2n + 2 + \text{Heart}$$

$$C_2H_4 + H_2 \xrightarrow{\text{Catalyst}} C_2H_6$$

This is know as Sabatier-senderen's reduction.

Platinum and palladium-black; *i.e.*, the metals in a very finely divided state, may be prepared by reducing their soluble salts with formaldehyde.

Adams' platinum-platinum oxide catalyst is prepared by reducing platinum oxide hydrogen before the addition of the compound being hydrogenated, or it may be added to the compound, reduction of the oxide taking place during hydrogenation.

Brown *et al.* (1962) have described the preparation of a 'highly active' platinum catalyst for alkene hydrogenation, and Eaborn *et al.* (1962) have prepared an even more active platinum catalyst.

One molecule of hydrogen is absorbed for each double bond present in the unsaturated compound.

The are of hydrogenation of olefinic bonds (at room temperature and atmospheric pressure) is:

$$— CH = CH_2 \; > \; — CH = CH— \text{ or a ring double bond.}$$

When the alkene is of the type R_2 C=CR_2 or R_2C = CHR, then hydrogenation under the above conditions is difficult.

The mechanism of catalytic hydrogenation is still not completely and erstood. It is widely accepted that the hydrogen is adsorbed on the metal surface and is present as hydrogen atoms ($H_2 \longrightarrow 2H$), and that the alkene is also adsorbed on the metal surface.

Although the nature of the bonding to the metal in both cases is uncertain, it appears that it is more chemical than physical, and so is described as a *chemisorphy bond.* Linstead *et al.* (1942), from their studies on catalytic hydrogenation, proposed that the less hindered side of an unsaturated molecule is adsorbed on the metal surface, and that this is then followed by the *simultaneous* addition of two hydrogen atoms.

In this way, the addition to form the *cis-product* was explained (attack must be from one side only). Furthermore, since chemisorption has converted hydrogen into hydrogen atoms and has broken one of the multiple bonds, the activation energy of catalytic hydrogenation is considerably lower than in the uncatalysed reaction.

Later work has shown that this mechanism is unsatisfactory. Isomerisation of alkenes may occur during hydrogenation, *e.g.,* Smith *et al.* (1962) isolated, in addition to n-butane, *cis*-and trans-but-2-ene from the products of the incomplete catalytic hydrogenation of but-1-ene. Also, although *cis*-addition usually predominated (*i.e.*, the addition is stereselective), *trans*-addition may occur and even predominate. These results and hydrogen-deuterium exchange experiments lead to the conclusion that addition of the adsorbed hydrogen atoms occurs one at a time and that the reaction is reversible (of catalytic dehydrogenation), *e.g.* (asterisk indicate metallic sites):

$$H_2 + CH_2 = CH_2\,H \rightleftharpoons \underset{\vdots}{H}\ \underset{\vdots}{H} + \underset{\vdots}{CH_2}—\underset{\vdots}{CH_2} \rightleftharpoons \underset{\vdots}{H} + \underset{\vdots}{CH_2}—CH_3$$

$$\rightleftharpoons CH_3—CH_3$$

When reducible functional groups are also present, *e.g.*, C = O, C ≡ N, CO_2R, etc., it is usually possible to find conditions to selectively reduce the olefinic bond (see text).

Hydrogenation of alkenes is also of interest from the thermodynamic point of view. Let us consider the hydrogenation of ethylene:

$$CH_2 = CH_2(g) \longrightarrow CH_3\ CH_3(g)$$

In this reaction, one C—C double bond and one H—H bond are broken (energy is supplied to the system) and one C— single bond and two C—H bonds are formed (energy is *given out by the system*) Hence, from bond-energy Table 2.2:

$\Delta H^{\oplus} = +\ (606.7 + 431.0) - (347.3 + 2 \times 414.2) = -\ 1138.0$ kJ (the observed value is $-$ 137.2kJ). Also, the entropy values *S*, 298 K) of ethylene, hydrogen and ethane are 219.7, 130.5, and 230.1 JK^{-1} respectively. $\therefore\ \Delta S^{\oplus} = 230.1 - (219.7 + 130.5) = -120.1 JK^{-1}$. Thus there is a *decrease* in entropy.

$$\Delta G^{\oplus} = \Delta H^{\oplus} - \Delta S^{\oplus} = -\ 138.0 - \ (298 \times -\ 120.1/1000)$$

$$= -\ 138.0 + 35.8 = -102.2 \text{ kJ}$$

Since $\Delta G^{\oplus}$ (at 25°C) has a large negative value, this means that the hydrogenation of ethylene is thermodynamically highly favourable. ($\Delta G^{\oplus} = -$ RT In K $= -2.3$ RT log K; $\therefore$ log K $= -\Delta G^{\oplus}/2.3\ RT = 102.2/5.7 = \sim 18$, *i.e.*, K $= 10^{18}$). Even so, the reaction does not proceed at any measurable rate at room temperature. As we have seen, the rate of a reaction depends on activation energies (and frequency factors). The activation energy of the above reaction is far too high to be attained at room temperature, but in the presence of a catalyst a new path with a much lower activation energy is now available, and so the reaction can therefore proceed at a fast rate.

The olefinic bond is not reduced by metal and acid, sodium and ethanol, lithium aluminium hydride, etc. unless it is a, b with respect to certain groups. *e.g.*, C = O.

Some chemical reagents, however, do reduce olefinic bonds, *e.g.*, alkenes are reduced to cis products by di-imide, and this stereospecificity can be explained by the formation of a cyclic T.S.

$$H-N=N-H + \underset{H\quad R}{\overset{H\quad R}{C=C}} \longrightarrow \left[\text{transition state}\right] \longrightarrow N\equiv N + \underset{H\quad H\quad R}{\overset{H\quad H\quad R}{C-C}}$$

Di-imide is an unstable solid (at low temperature), and so is prepared *in site*, *e.g.*, by the oxidation hydrazine with hydrogen peroxide, etc. Di-imide is selective in that it reduces carbon—carbon and nitrogen—nitrogen multiple bonds, but does not usually reduce $C = O$, NO_2, $C\equiv N$, etc.

Reduction by di-imide is an example of *transfer hydrogenation* (hydrogen is supplied by a donor molecule which is itself oxidised). Another donor molecule is cyclohexene (which is oxidised to benzene):

$$\text{cyclohexene} + 2\ \rangle C=C\langle \xrightarrow[100\ C]{Pd} \text{benzene} + 2\ \rangle CH-CH\langle$$

This system reduces many types of functional groups, *e.g.*, olefinic and acetylenic bonds, NO_2,. — N = N —, etc.

A *terminal* double bond may be reduced by sodium in liquid ammonia in the presence of an alcohol (MeOH or EtOH; alcohols are stronger acids than ammonia). This method is known as the Birch reduction, and is believed to proceed step wise via an anionic free radical:

$$RCH = CH_2 \xrightarrow[e]{Na} R\dot{C}H\bar{C}H_2 \xrightarrow{ErOH} R\dot{C}HCH_3 \xrightarrow[e]{Na}$$

$$R\bar{C}HCH_3 \xrightarrow{ErOH} RCH_2CH_3$$

The double bond is also reduced in excellent yield by $NaBH_4$ — $PtCl_2$ (Brown *et al.*, 1962).

Addition of halogen : Alkenes react with halogens to form dihaloalkanes. The order of reactivity is chlorine > bromine > iodine.

Alkenes form addition compounds with chlorine or bromine, *e.g.*, ethylene adds bromine to form ethylene dibromide:

$$CH_2 = CH_2 + Br_2 \longrightarrow BrCH_2CH_2Br \quad (85\%)$$

Hetrolytic (polar) mechanise : Halogen addition can take place either by a heterolytic (polar) or a free-radical mechanism.

Heterolytic (polar) mechanism. Halogen addition readily occurs in solution, in the absence of light or peroxides, and is catalysed by inorganic halides, *e.g.*, aluminium chloride, or by polar surfaces (of the reaction vessel). These facts lead to the conclusion that reaction occurs by a polar mechanism. Further evidence for the polar mechanism comes from the work of Francis (1925), who showed that when ethylene reacts with bromine in aqueous sodium chloride solution, the products are ethylene dibromide and 1-bromo-2-chloroethane; *no ethylene dichloride is obtained.*

$$Br_2 + CH_2 = CH_2 \xrightarrow[H_2O]{NaCl} BrCH_2CH_2Br + BrCH_2CH_2Cl$$

Francise also Showed

$$Br_2 + CH_2{=}CH_2 \xrightarrow{aq.\ NaI} BrCH_2CH_2Br + BrCH_2CH_2Cl$$

$$Br_2 + CH_2{=}CH_2 \xrightarrow{aq.\ NaNO_3} BrCH_2CH_2Br + BrCH_2CH_2CH_2ONO_2$$

This work demonstrates that these addition reactions are at least a two-stage process, but does not demonstrate the order of the addition, *i.e.*, whether halogens are electrophilic or nucleophile reagents. However, coupled with the fact that ethylene does not react with these solutions in the absence of bromine, this is a strong indication for electrophilic attack by the halogen. A mechanism consistent with these facts is:

$$CH_2{=}CH_2 \quad Br{-}Br \xrightarrow{slow} \overset{+}{C}H_2{-}CH_2Br + Br^-$$

$$BrCH_2CH_2Br \xleftarrow[fast]{Br^-} \quad \xrightarrow[fast]{Cl^-} ClCH_2CH_2Br$$

In addition to the evidence described above for electrophilic attack by halogens, there is the argument that negative ions (nucleophilic reagents) are hindered from attacking ethylenic carbon atoms by the screen of π-electrons. In fact, π-electrons are very susceptible to attack by electrophilic reagents. That alkenes are themselves nucleophilic reagents; *e.g.*, many form addition complexes with the silver ion as *e.g.*, perchlorate. These addition compounds are known as π-complexes; the silver atom is not directly joined to either carbon atom, but is bound by overlap of the π-orbital with the vacant silver ion orbital.

$$Me_2C{=}CMe_2 + Ag^+ \longrightarrow \left.\underset{\downarrow\atop Ag^+}{Me_2C{=}CMe_2}\right\} ClO_4$$

Stereochemistry of halogen addition to alkenes (and alkynes) : Addition of halogen is usually predominantly trans, *i.e.*, the addition is stereoselective (one stereoisomer predominates over the other), *e.g.*, the addition of bromine to maleie acid gives (±)-dibromosuccinic acid, which can result only from *trans* addition. To account for this, Roberts *et al.* (1937) suggested that the first step is the formation of a bridged (or *non-classical*) carbonium ion, followed by attack by the bromide ion.

H H C=C HO_2C CO_2H $\xrightarrow{Br_2}$ Br^+ H HO_2C H HO_2H $\xrightarrow{Br_2}$

H CO_2H Br Br H CO_2H (I) + Br CO_2H H CO_2H Br (II)

(±)

The bromide ion attacks from *behind and along* the bonding C — Br+ line, and this results in Walden inversion at this carbon atom. Since the *brominium* (*bromonium*) ion is symmetrical, each carbon atom is attacked equally well, and consequently results in the formation of (I) and (II) in *equal* amounts. (I) and (II) are a pair of enantiomers and hence the product is a racemic modification.

The existence of this bridged ion has been demonstrated by various workers. I should be noted, however, that although addition is usually *trans, cis* addition may also occur in certain circumstances.

Free-radical mechanism : It has generally been accepted that addition of halogen to alkenes, in the *absence of light,* is polar. Thus, for example, Stewart *et at.* (1935) showed that the addition of chlorine to ethylene is accelerated by light. This suggests a free-radical mechanism:

$$Cl_2 \xrightarrow{Cl_2} 2Cl$$

$$CH_2 = CH_2 + Cl \longrightarrow CH_2Cl{-}CH_2 \xrightarrow{Cl_2} CH_2Cl{-}CH_2Cl; \text{ etc.}$$

Poutsma (1965), however, has shown that chlorine adds, *e.g.*, to but-2-ene at 25°C in the *absence* of light, and proposed a *spontaneous* free-radical mechanism:

$$MeCH = CHMe + Cl_2 \xrightarrow{\text{spont}} MeCHClCHMe$$

$$\xrightarrow{Cl_2} MeCHClCHClMe + Cl; \text{etc.}$$

This mechanism is supported by the fact that the reaction was much slower in the presence of oxygen.

Under these conditions, free-radical addition is inhibited and the reaction proceeds by the slower polar mechanism.

Instead of addition reactions with the halogens, the alkenes may undergo substitution provided the right conditions are used. Thus, when straight-chain alkenes are treated with chlorine at a high temperature, they form mainly monochlorides of the allyl type *i.e.*, in the chain — C — C — C = C —, it is the hydrogen of C that is substituted. This high temperature substitution is strongly presumptive of a free-radical mechanism, *e.g.*,

$$Cl_2 \xrightarrow{400^\circ - 600^\circ C} 2Cl\cdot$$

$$CH_3CH = CH_2 + Cl\cdot \longrightarrow HCl + \cdot CH_2 = CH = CH_2$$

$$\xrightarrow{Cl_2} ClCH_2CH = CH_2 + Cl\cdot;\ \text{etc.}$$

The reason for allyl substitution occurring is not certain, but one possibility is that, of the three types of free radicals that may be formed, $CH_3CH = CH\cdot$, $CH_3\dot{C} = CH_2$, and $\cdot CH_2CH = CH_2$, it is the allyl one that will have the greatest resonance stabilisation (the contributing structures are identical):

$$\cdot CH_2CH = CH_2 \leftrightarrow CH_2 = CHCH_2\cdot$$

From the M.O. point of view, in the allyl free radical, each carbon atom is associated with a p_t electron. Thus an M.O. is formed covering all three carbon atoms, with consequent delocalisation. Since the formation of the free radical with maximum stability means that the reaction path through this intermediate will be accompanied by the greatest decrease in free energy, the reaction will therefore proceed along this route.

An alternative explanation is that of the two types of C—H bonds present, viz., —C—H and = C—H, the bond energy of the latter is greater than that of the former, and so the methyl group will be attacked preferentially.

Allyl substitution in alkenes is also readily carried out with N-bromosuccinimide.

The action of fluorine on alkenes usually results in the formation of carbon tetraflouride, but addition to the double bond may be effected by treating the alkene with hydrogen fluoride in the presence of lead dioxide (Henne et al., 1945); the fluorinating agent is lead tetraflouride:

$$PbO_2 + 4\,HF \longrightarrow PbF_4 + 2H_2O$$

$$\rangle C{=}C\langle \; + PbF_4 \longrightarrow \; \rangle CFCF\langle \; + PbF_2$$

Oxidation of alkenes is readily effected in the allyl position by selenium dioxide, *e.g.*,

$$MeCH_2CH = CHMe \xrightarrow{SeO_2} MeCHOHCH = CHMe$$

It should be noted that CH2 group is preferred to the 1— Me group, Alkenes of the type $R^1CH{=}CHR^2$ are oxidised to α-diketones, R^1cocoR^2.

Alkenes form addition compounds with the halogen acids, *e.g.*, ethylene adds hydrogen bromide to form ethyl bromide:

$$C_2H_4 + HBr \longrightarrow C_2H_5Br$$

The order of reactivity of the addition of the halogen acids is hydrogen iodide > hydrogen bromide > hydrogen chloride > hydrogen fluoride (this is also the order of acid strength). The conditions for the addition are similar to those for the halogens; the addition of hydrogen fluoride, however, is effected only under pressure..

In the case of unsymmetrical alkenes it is possible for the addition of the halogen acid to take place in two different ways, *e.g.*, propene might add on hydrogen iodide to form propyl iodide,

$$CH_3CH = CH_2 + HI \longrightarrow CH_3CH_2CH_2I$$

or it might form isopropyl iodide:

$$CH_3CH = CH_2 + HI \longrightarrow CH_3CHICH_3$$

Markownik off studied many reactions of this kind, and as a result of his work, formulated the following rule: *the negative part of the addendum adds on to the carbon atom that is joined to the least number of hydrogen atoms.* In the case of the halogen acids the halogen atom is the negative part, and so isopropyl halide is obtained.

Markownik off's rule is empirical, but may be explained theoretically on the basis that the addition occurs by a polar mechanism. As with halogens, the addition of halogen acid is an electrophilic reaction, the proton adding first, followed by the halide ion. Also, the addition is predominantly *trans,* and this may be explained in terms of the formation of a bridged carbonium ion :

$$CH_2{=}CH_2 \quad H{-}Cl \xrightarrow{slow} CH_2\overset{H}{\underset{+}{\triangle}}CH_2 + Cl^- \xrightarrow{fast} CH_3 + CH_2Cl$$

Now consider the case of propene. Since the methyl group has a + I effect, the α-electrons are displaced towards the terminal carbon atom which, in consequence, acquires a negative charge. Thus, the proton adds on to the carbon *farthest* from the methyl group, and the halide ion then adds to the carbonium ion:

$$CH_3 \longrightarrow CH = CH_2 \quad H{-}I \rightleftharpoons CH_3\overset{+}{C}HCH_3 + I^- \longrightarrow CH_3CHICH_3$$

An alternative explanation for Markownik off's rule is in terms of stabilities of carbonium ions (I), (II) and (III) represent a primary, secondary and tertiary carbonium ion, respectively.

$$\underset{(I)}{MeC\overset{+}{H_2}} \quad \underset{(II)}{Me_2CH^+} \quad \underset{(III)}{Me_3C^+} \quad \underset{(IV)}{\overset{\delta}{Me} \longrightarrow \overset{\Delta}{C}H \longleftarrow \overset{\delta}{Me}}$$

The positive charge on the central carbon atom makes this atom strongly electron-attracting, and this direction of electron movement is assisted by the + I effect of the Me group, and consequently the charge on the central carbon atom is partially neutralised (represented by Δ + in IV) and, at the same time, Me acquires a small positive charge. Thus the net result is spreading of the charge (as in IV), and this produces increased stability. The reaction will therefore proceed through the intermediate formation of the ion.

Still another explanation for the order of the stability of the carbonium ions is *no bond resonance*, *e.g.*,

$$H{-}\underset{H}{\overset{H}{C}}{-}\underset{H}{\overset{H}{C^+}} \longleftrightarrow H{-}\underset{H}{\overset{H^+}{\bar{C}}}{-}\underset{H}{\overset{H}{C^+}} \longleftrightarrow H{-}\underset{H}{\overset{H^+}{C}}{=}\underset{H}{\overset{H}{C}} \longleftrightarrow$$

Thus the number of hydrogen atoms available for no bond resonance is 3 for (I), 6 for (II) and 9 for (III). Consequently, (III) would be expected to be the most stable. At the same time, the charge is spread (over the methyl hydrogens).

The direction of addition of ICI and BrCI to propene is explained in the same way:

$$CH_3CHClCH_2I \xleftarrow{ICl} CH_3CH = CH_2 \xleftarrow{BrCl} CH_3CHClCH_2Br$$

Chlorine is more electronegative than either iodine or bromine, and so is the negative part of the addendum and adds in the *second step.*

Any factor that lowers the activation energy of a reaction will increase the rate of that reaction. Hence, the more stable the carbonium ion the faster will be the additional. Since the order of stability is (III) > (II) > (I), it can be anticipated that the rate of addition of HX will be

$$Me_2C = CH_2 > MeCH = CH_2 > CH_2 = CH_2$$

This order is observed in practice.

In certain cases (according to the nature of the carbonium ion), these addition reactions may be accompanied by rearrangement.

Relative reactivities of various positions in a molecule. The addition of halogen acid to unsymmetrical alkenes is an example of the prediction of the relative reactivities of various positions in a given molecule. Two different methods were used, and these are widely applied in the study of reaction mechanisms.

(i) *Predictions based on charge distribution.* This approach assumes that the reactivity of a position in a polar reaction is determined by the charge density at that position, *i.e.*, an electrophilic reagent will attack preferentially at the site of highest electron density and a nucleophilic reagent will attack at the site of lowest electron density. The charge distribution is ascertained by the examination of the consequences of inductive and/or resonance effects. This approach is referred to as the isolated molecule method. Two disadvantages are the neglect of changes in charge distribution as the reagent approaches the substrate and the neglect of electrostatic forces between the substrate molecule and a charged reagent. Finar (1968) has shown that these electrostatic forces may be the predominant factor in predicting the site of attack.

(ii) *Predictions based on stabilities of intermediates.* This method assumes that the reactivities of different positions in a given molecule are determined only by their relative activation energies. Since the activation energy depends on the structure of the transition state, a difficulty here is the nature of the model that is to be used for the T.S. For reactions that are believed to involve intermediate carbonium ions, carbanions and free radicals, it is these that are chosen to represent the structures of the transition states, the stabilities of which are then examined in terms of inductive and/or resonance effects. The assumption is then made that the more stable the inter mediate (*i.e.*, the lower its energy), the lower is the energy of the T.S., and consequently the lower is the energy of activation leading to that intermediate. This approach is referred to as the localisation energy method. It also has disadvantages, *e.g.*, electrostatic forces (*cf.* above) and the nature of the reagent are neglected. Since the latter factor plays a great part in deciding the structure of the T.S ., the use of the intermediate as the model for all types of reagents may lead to wrong predictions.

In this book, many reaction mechanisms are dealt with by both methods, since it is instructive to compare them.

The Peroxide Effect (Kharasch, 1933). The presence of oxygen or peroxides that are formed when the alkene stands exposed to the air, or added peroxides such as benzoyl peroxide, causes the addition of hydrogen bromide to take place in the direction opposite to that predicted by Markownik off's rule. This departure from the rule is known as the 'abnormal' reaction, and was shown to be due to the 'peroxide effect' (Kharasch *et al.*, 1933). Hydrogen chloride, hydrogen iodide and hydrogen fluoride do not exhibit the abnormal reaction. It has been found that the addition of hydrogen bromide is 'abnormally' effected photochemically as well as by peroxide catalysts.

The mechanism of the peroxide effect is a free-radical chain reaction, the peroxide generating the free radical $R^1\cdot$ (*cf.* polymerisation, below):

$$(R^1CO_2)_2 \longrightarrow 2R^1CO_2\cdot \longrightarrow 2R^1\cdot + 2CO_2$$

$$R^1\cdot + HBr \longrightarrow R^1H + Br\cdot$$

$$R^1CH = CH_2 + Br \longrightarrow R^2CHCH_2Br \xrightarrow{HBr} R^2CH_2CH_2Br + Br, \text{ etc.}$$

In the photochemical addition, the bromine atom is produced by absorption of a quantum of light:

$$HBr \xrightarrow{hv} H + Br.$$

Abell et al. (1962), using ESR spectroscopy, showed that the photochemical of HBr produces free radicals which appear to be best represented as bridged structures. They also showed, using DBr that the bromine atom is the initial event in the attack.

The reason for the addition being contrary to Markownik off's rule is not certain. A favoured theory is that the order of stability of free radicals is the same as that for carbonium ions, *i.e.*, t > s > primary. One explanation for this order is no bond resonance. Thus the reaction proceeds as shown in the equations above, the primary free-radical. $R^2CHBrCH_2^{\cdot}$, being much less favoured energetically than the secondary. R^2CHCH_2Br.

$$H{-}\underset{H}{\overset{H}{C}}{-}\underset{H}{\overset{H}{C}} \longleftrightarrow H{-}\underset{H}{\overset{H^+}{C}}{-}\underset{H}{\overset{H}{C^+}} \longleftrightarrow H{-}\underset{H}{\overset{H}{C}}{=}\underset{H}{\overset{H}{C}} \longleftrightarrow \cdots\cdots$$

It has previously been pointed out that, among other factors, the energy of activation in a given reaction is lower the greater the strength of the new bond formed. Now, the bond broken is the y-bond, and so this energy change will not affect the activation energy whichever way the free-radical addition occurs. Also, the strength of a C—H bond is in the order p > s > i, this order also applies to bonds other than C—H. Hence, on this basis, it can be expected that the intermediate free radical will be preferably $RCHCH_2X$ rather than R $CHXCH_2^{\cdot}$. This then is another possible explanation for free-radical addition occurring contrary to Markownik off's's rule.

The abnormal reaction in the presence of peroxides can be prevented by the addition, in sufficient amount, of an *inhibitor* such as diphenylamine, catechol, etc. Inhibitors combine with free radicals, and so prevent propagation of the chain-reaction. Thus an inhibitor produces (if present in small amount only) an induction period, *i.e.*, a period when apparently no reaction is occurring. When the inhibitor is used up, it is then possible for the chain reaction to proceed in the normal fashion.

Another problem is why only hydrogen bromide exhibits the peroxide effect. A reasonable explanation may be offered in terms of the energetics of the chain steps in radical additions of HX to ethylene (kl mol^{-1} at 25°C).

$\Delta H_1^{\oplus}$ H—X	$\Delta H_2^{\oplus}$ $X + CH_2 = CH_2 \rightarrow$ $XCH_2{-}CH_2$	 $XCH_2{-}CH_2 + H{-}X \rightarrow$ $XCH_2{-}CH_3 + X$
H—Cl	–67.0	+ 12.6
H—Br	–25.1	–50.2
H—I	–46.0	–117.1

If we assume that $\Delta S^{\oplus}$ is the same for the corresponding steps in the three reactions, then — RT In $K = \Delta G^{\oplus} = \Delta G^{\oplus}$ + constant. Thus,

$\Delta G^{\oplus}$ may be equated with $\Delta H^{\oplus}$, and hence the more negative ΔH^{θ} is, the more thermodynamically favourable will be the reaction. Only for HBr are both steps exothermic and consequently only for HBr is each step thermodynamically favourable.

$$H + CH_2 = CH_2 \longrightarrow CH_3{-}CH_2 \xrightarrow{HBr} CH_3CH_2Br + H$$

$$\Delta H^{\oplus} = -\ 167.4\ kJ \qquad \Delta H^{\oplus} = +\ 100.4\ kJ$$

Although the first step is highly the modynamically, favourable, the reverse is true for the second step. Hence, the free-radical addition of HBr occurs first by addition of Br·, followed by addition of H·.

Tri-and tetra-halogenated methanes also add on to a *terminal* double bond in the presence of peroxides (Kharasch et al., 1945-1950). Here again the mechanism is believed to be a free-radical chain reaction. *e.g.*,

$$(RCO_2)_2 \longrightarrow 2R{\cdot}+2CO_2$$

$$R{\cdot}+CHCl_3 \longrightarrow RH+{\cdot}CCl_3$$

$$RCH{=}CH_2+{\cdot}CCl_3 \longrightarrow RCHCH_2CCl_3$$

$$\xrightarrow{CHCl_3} RCH_2CH_2CCl_3+{\cdot}CCl_3;\ etc.$$

Hypohalous acids add on to alkenes to form halohydrins. Usually the reaction is carried out by treating the alkene with chlorine or bromine-water; alkene dihalide is also formed.

The mechanism is believed to be

$$CH_2 = CH_2+Br_2 \rightleftharpoons Br^-+Br\overset{+}{C}H_2CH_2 \xrightarrow{H_2O} BrCH_2CH_2\overset{+}{O}H_2$$

$$\longrightarrow H^+ + Br\,CH_2CH_2\,OH$$

With unsymmetrical alkenes, the unsymmetrical alkenes, the hydroxyl group adds on to the carbon atom joined to the least number of hydrogen atoms:

$$RCH{=}CH_2\ \ Cl{-}Cl \longrightarrow Cl^- + R\overset{+}{C}HCH_2Cl \xrightarrow{H_2O} RCHOHCH_2Cl + H^+$$

Aqueous solutions of hypohalous acid also add on to alkenes in the presence of strong acids to form halohydrins *e.g.*, the addition of hypochlorous acid to ethylene. The mechanism is probably via the formation of the *chloronium ion:*

$$ClOH + H^+ \rightleftharpoons Cl\overset{+}{O}H_2 \rightleftharpoons Cl^+ + H_2O$$

$$CH_2 = CH_2 + Cl^+ \longrightarrow ClCH_2\overset{+}{C}H_2 \xrightarrow{H_2O} ClCH_2CH_2OH_2^+$$

$$\underset{-H^+}{\rightleftharpoons} ClCH_2CH_2OH$$

Halohydrins regenerate the alkene when treated with zinc dust in acetic acid (*ef.*)

The addition of 'HOX' involves the addition of the solvent, in this case , water. Water is a nucleophilic solvent and competes with the other nucleophilic, Cl^-, which is present. All other nucleophilic solvents behave in the same way as water, *e.g.*, propene reacts with methanol in the presence of sulphuric acid at follows:

$$MeCH{=}CH_2 \xrightarrow{H_2SO_4} Me_2CH^+ \xrightarrow{MeOH} Me_2CH\overset{+}{O}(H)Me \xrightarrow{-H^+} Me_2CHOMe$$

Various sulphenyl halides form adducts with alkenes; 2, 4-dinitrobenzenesulphenyl chloride in particular has been found extremely useful for identifying alkenes (Karasch *et* al., 1949-1952).

$$Ar\ddot{S}{-}\ddot{C}l + CH_2{=}CH_2 \longrightarrow CH_2{-}CH_2 \text{ (bridged by } \overset{+}{S}Ar) + Cl^- \longrightarrow ArSCH_2CH_2Cl$$

Two products are obtained with unsymmetrical alkenes, the predominating adduct being accepted as the one formed in accordance with Markownik off's's rule. However, Calo *et* al. (1968) have shown that the direction of addition depends on the nature of the solvent.

Compounds containing triple bonds also form adducts with one molecule of the sulphenyl chloride.

Alkenes are absorbed by concentrated sulphuric acid to form alkyl hydrogen sulphates. Addition takes place according to Markownik off's's rule, *e.g.*, propene reacts with sulphuric acid to form isopropyl hydrogen sulphate;

$$CH_3CH{=}CH_2 \quad H{-}OSO_2OH \longrightarrow CH_3\overset{+}{C}HCH_3 + \overset{-}{O}SO_2OH$$

$$\longrightarrow (CH_3)_2CHOS_2OH$$

Alkanes are not absorbed by cold concentrated sulphuric acid, and hence may be separated form alkenes (see also ethers and alcohols).

Hydrogen of alkenes. Alkenes may be hydrated to alcohols by absorption in concentrated sulphuric acid followed by hydrolysis of the alkyl sulphate (see above section and). Alkenes, however, may also be catalytically hydrated in dilute acid solution, *e.g.* Lucas *et al.* (1934) showed that isobutene forms t-butanol in dilute acid solution:

$$(CH_3)_2C = CH_2 + H_2O \xrightarrow{H^+} (CH_3)_3COH$$

The mechanism of alkene hydration has been the subject of much discussion. The hydration reaction, as such, cannot be examined kinetically, but since the reaction is reversible, the principle of microscopic reversibility may be applied. Hughes and Ingold (1941) have examined the elimination of water from t-alcohols, and on the basis of their results have proposed the following mechanism for hydration.

$$Me_2C{=}CH_2 \quad H{-}\overset{+}{O}H_2 \xrightleftharpoons{\text{slow}} Me_2\overset{+}{C}{-}Me + H_2O$$

$$\rightleftharpoons Me_2\underset{{}^{+}OH_2}{\underset{|}{C}}{-}Me \xrightleftharpoons{-H^+} Me_2\underset{OH}{\underset{|}{C}}Me$$

Since hydration cannot be carried out in alkaline solution, this is very strong evidence that the first step is proton addition in acid-catalysed hydration, *i.e.*, the reaction is electrophilic. Further support for this mechanism is the fact that only those alkenes which form t-carbonium ions are readily hydrated under these conditions. Such ions are more stable than s-or prim.-carbonium ions and consequently will be produced more readily than the latter.

It appears the hydration is not stereoselective, as is the case for halogen acid addition (Collins *et al.*, 1960). This may be explained on the basis that a classical carbonium ion is formed (as shown in the mechanism).

A further point of interest is that water is not a sufficiently strong acid to transfer a proton to an alkene; hence the use of dilute acid solution, In concentrated acid, the nucleophilic an ion of the acid adds. In sufficiently diluted acid, $[H_2O] >> [HSO_4^-]$ and consequently water is involved in the second step.

Since hydration of alkenes and dehydration of alcohols are reversible, both proceed through the same intermediate carbonium ions, and if these are capable of undergoing rearrangement, 'unexpected' alcohols may be obtained from the alkene. Rearrangement, however, may be avoided by treatment of the alkene with mercuric acetate (in tetrahydrofuran). followed by reduction in aqueous sodium borohydride (Brown *et al.*, 1967).

$$Me_3CCH{=}CH_2 \xrightarrow[\text{THF}]{Hg(OOCMe)_2} Me_3CC\underset{OOCMe}{\underset{|}{H}}CH_2HgOOCH_2HgOOCMe$$

$$\xrightarrow[\text{NaOH}]{NaBH_2} Me_3CCHOHMe + Hg$$

This method is referred to as oxymercuration-demercuration, and the final result is hydration in accordance with Markownik off's's rule (*cf.* hydroboration, below).

Alkenes add on nitrosyl chloride, nitrosyl bromide and oxides of nitrogen: *e.g.,* ethylene nitrosochloride with nitrosyl chloride:

$$CH_2 = CH_2 + NOCl \longrightarrow ClCH_2CH_2NO$$

Since the X atom is the negative end of the dipole in NOX, it will add on to the carbon atom joined to the least number of hydrogen atoms (Markownik off's's rule), *e.g.,* trimethylethylene adds on nitrosyl bromide to form the following trimethylethylene nitrosobromide.

$$(CH_3)_2C = CHCH_3 + NO{-}Br \longrightarrow (CH_3)_2CBrCH(NO)CH_3$$

The addition of oxides of nitrogen to alkenes is complicated, the products formed depending on the structure of the alkene and the nature of the 'nitrous fumes'.

Alkenes are readily hydroxylated, *e.g.,* add on hydroxyl groups, to form dihydroxy compounds known as glycols (*q.r.*), Hydroxylation may be effected:

(i) By cold dilute alkaline permanganate solution (*cis*-hydroxylation): *e.g.,* ethylene is converted into ethylene glycol:

$$CH_2 = CH_2 + H_2O + [O] \xrightarrow{KMnO_4} HOCH_2CH_2OH$$

The mechanism of hydroxylation with permanganate is believed to proceed via cyclic intermediate (*cf.* osmium tetroxide, below).

$$\begin{matrix} RCH \\ \| \\ RCH \end{matrix} \xrightarrow{MnO_4} \begin{matrix} RCH{-}O \quad O \\ | \quad \diagdown Me \diagup \\ RCH{-}O \quad O \end{matrix} \xrightarrow{OH} \begin{matrix} RCH{-}O \\ | \\ RCH{-}O{-}OMnO_3H^- \end{matrix}$$

$$\xrightarrow{OH} \begin{matrix} RCH{-}OH \\ | \\ RCH{-}OH \end{matrix}$$

The mechanism accounts for cis-hydroxylation and is supported by the work of Wiberg *et al. (1957)*, who used potassium permanganate labelled with ^{18}O and allowed that both glycol oxygen atoms come from the oxidising agent.

Similarly, chromic acid converts alkenes into glycols (or epoxides). The mechanism again involves a cyclic intermediate ($HCrO_4^-$ replaces MnO_4^-, and Cr $(O^-)OH$ replaces Mn $(=O)O^-$)

(ii) By 90 percent hydrogen peroxide in glacial acetic acid or better, in formic acid; *e.g.*, oleic acid is converted into 9,10-dihydroxystearic acid:

$$CH_3(CH_2)_7CH = CH(CH_2)_7CO_2H + H_2O_2 \longrightarrow CH_3(CH_2)_7CH(OH)CH(OH)(CH_2)_7CO_2H$$

The addition of hydrogen peroxide may be catalysed by various oxides, *e.g.*, osmium tetroxide in t-butanol (*cis*-addition), selenium dioxide in t-butanol or acetone (trans-addition;)

(iii) By means of osmium tetroxide. This compound adds very readily to an ethylenic double bond at room temperature (Criegee, 1936):

$$\begin{matrix} RCH \\ \| \\ RCH \end{matrix} + OsO_4 \longrightarrow \begin{matrix} RCH-O \\ | \\ RCH-O \end{matrix} \!\!\!\!> Os \begin{matrix} \nearrow O \\ \searrow O \end{matrix} \quad (r.g.\text{-}ex)$$

These cyclic compounds (osmic esters), on refluxing with aqueous ethanolic sodium hydrogen sulphite, are hydrolysed to 1,2-glycols (*cis*-glycols).

If the addition of osmium tetroxide is carried out in the presence of pyridine, coloured crystalline compounds are obtained, usually in theoretical yield;

$$\begin{matrix} RCH \\ \| \\ RCH \end{matrix} + OsO_4 + 2C_3H_5N \longrightarrow \begin{matrix} RCH-O \\ | \\ RCH-O \end{matrix} \!\!\!\!> OsO_2, 2C_5H_5M$$

Berkowitz *et al.* (1958) have shown that ruthenium tetroxide is more convenient to use than osmium tetroxide; it is less toxic.

cis-Hydroxylation of a double bond may also be effected by treating the alkene with iodine and silver acetate in wet acetic acid, and then hydrolysing the mixed mono- and di-acetates with alkali (Barkley *et al.*, 1954):

$$>C=C< \xrightarrow[CH_3CO_2Ag]{O_2} \begin{matrix} >C=C< \\ OH \quad OCOCH_3 \end{matrix} + \begin{matrix} >C=C< \\ CH_3COO \quad OCOCH_3 \end{matrix} \xrightarrow{NaOH} \begin{matrix} >C=C< \\ OH \quad OH \end{matrix}$$

If the above reaction is carried out in dry acetic acid, the di-acetate and diol have the trans-configuration.

Glycols (or the alkenes themselves) are readily oxidised to acids or ketones by means of acid permanganate or acid dichromate, the nature of the products being determined by the structure of the glycol, *e.g.*,

(a) propene glycol gives acetic and formic acid:

$$CH_3CH(OH)CH_2OH \xrightarrow{[O]} CH_3CO_2H + HCO_2H$$

(b) Isobutene glycol gives acetone and formic acid:

$$(CH_3)_2C(OH)CH_2OH \xrightarrow{[O]} (CH_3)_2CO + HCO_2H + H_2O$$

Oxidation of a glycol may also be effected by lead tetra-acetate, $(CH_3CH_2)_4$ Pb, or by periodic acid, HIO_4 or H_5IO_6, the products being aldehydes or ketones, according to the structure of the glycol, *e.g.*,

(a) ethylene glycol gives two molecules of formaldehyde:

$$HOCH_2CH_2OH \xrightarrow[(CH_3CO_2)_4Pb]{[O]} HCHO + HCHO$$

(b) Isobutene glycol gives acetone and formaldehyde:

$$(CH_3)_2C(OH)CH_2OH \xrightarrow[HIO_4]{[O]} (CH_3)_2CO + HCHO$$

It can be seen that whatever oxidising agent is used the glycol is split into two fragments, the rupture of the carbon chain occurring between the two carbon atoms joined to the hydroxyl groups. Since these two carbon atoms were linked together by the double bond in the original alkene, identification of the two fragments which may be acids, aldehydes or ketones, will indicate the position of the double bond in the alkene, *e.g.*,

$$R^1CH{=}CHR^2 \xrightarrow{H_2O_2} R^1CH(OH)CH(OH)R^2 \xrightarrow{HIO_4} R^1CHO + R^2CHO$$

The method of oxidative cleavage of the olefinic bond described above has now been replaced by the use of the Lemieux reagent (1955). This is better both for determining the position of a double bond and for preparing carbonyl compounds. The reagent is an aqueous solution of sodium periodate and a trace of potassium permanganate. The alkene is oxidised to the *cis*-diol, which is then cleaved by the periodate to aldehydes and/or ketones. Aldehydes are further oxidised by the permanganate to acids (formaldehyde is usually obtained from a terminal alkene). The reaction proceeds at room temperature, and the lower valent form of Mn is reoxidised to the higher valent form by the periodate (hence only a catalytic trace of permanganate is needed).

Instead of permanganate, a trace of osmium tetroxide has also been used by Lemieux *et al.* In this case, aldehyde is produced (it is not oxidised further), *e.g.*,

$$\begin{matrix} RCH \\ \| \\ RCH \end{matrix} + OsO_4 \longrightarrow \begin{matrix} RCH-O \\ | \\ RCH-O \end{matrix}\!\!\!>OsO_2 \xrightarrow{2NaHO_4} \begin{matrix} RCH \\ \| \\ RCH \end{matrix} + OsO_4 + 2NaIO_3$$

Prileschaieve's reaction, (1912). By means of per-acids, the double bond in alkenes is converted into the *epoxide* (alkene oxide). Perbenzoic acid, $C_6H_5COO_2H$, monoperphthalic acid $HO_2CC_6H_4COO_2H$, and p-nitroperbenzoic acid, have been used; *e.g.*,

$$R^1CH{=}CHR_2 + C_6H_5COO_2NH \longrightarrow R^1\overset{\diagup O \diagdown}{CH{-}CHR^2} + C_6H_5CO_2NH$$

Emmons *et al. (1954,* 1955) have found that peroxytrifluoroacetic acid (CF_3COO_2H) is a very good reagent for epoxidation and hydroxylation.

$$RCH{=}CHR_2 \xrightarrow{CF_3CO_3H} R\overset{\diagup O \diagdown}{CH{-}CHR} \xrightarrow{CF_3CO_3H}$$

$$RCHOHCH(OCOCF_3)R \xrightarrow[CH_3OH]{HCl\ in} RCHOHCHOHR \quad (60\text{-}95\%)$$

This method is particularly useful for high molecular-weight alkenes with a terminal double bond (these are only slowly hydroxtlated by other per-acid). Furthermore, peroxytrifluoroacetic acid may be used to hydroxylate α β-unsaturated carbonyl compounds, *e.g.*, ethyl acrylate, $CH_2{=}CHCO_2C_2H_5$.

Many mechanisms have been proposed for epoxidation, but none is certain. According to Pausacker *et al.* (1955), the mechanism is:

$$\begin{matrix} | \\ CH \\ \| \\ CH \\ | \end{matrix} + O\begin{matrix} \diagup O-\overset{}{C}C_6H_5 \\ \diagdown H \quad \overset{\|}{O} \end{matrix} \longrightarrow \underset{\text{trasnition state}}{\begin{matrix} | \\ CH \\ \vdots \\ CH \\ | \end{matrix} \cdots O \begin{matrix} O-CC_6H_5 \\ H-O \end{matrix}} \longrightarrow \begin{matrix} | \\ CH \\ | \\ CH \\ | \end{matrix}\!\!>O + \begin{matrix} O{=}CC_6H_5 \\ | \\ H-O \end{matrix}$$

This mechanism accounts for the fact that epoxidation is usually stereopecific, *i.e.*, one product a formed from each geometrical isomer.

Furthermore, the diol produced has the trans configuration, and this is in keeping with attack on the intermediate epoxide

Alkenes add ozone to form *ozonides*. These are usually explosive in the free state, and their structure and mechanism of formation have been the subject of a great deal of work. Staudinger (1922) suggested that the *molozonide* is formed first, and this then rearranges to the ozonide, some dimerising (to the diperoxide); polymers may also be formed. A widely accepted mechanism involves a 1,3-dipolar cyclo addition reaction (p. 394; this accounts for the stereospecific *cis*-addition to form the molozonide); ozone is considered to behave as:

$$O{=}\overset{+}{O}{-}\overset{-}{O} \longleftrightarrow \overset{+}{O}{-}O{-}\overset{-}{O}:$$

$$RCH{=}CHR \xrightarrow{O_3} \underset{\text{molozonide}}{RCH(O{-}O{-}O)CHR} \longrightarrow \left[\underset{\text{dipolar ion}}{RCH{=}O + \overset{-}{O}{-}O{-}\overset{+}{C}HR}\right]$$

$$\longrightarrow \underset{\text{ozonide}}{RCH(O{-}O)(O)CHR} + \underset{\text{diperoxide}}{RCH(O{-}O)(O)CHR}$$

This mechanism of fission followed by recombination is supported by the fact that if the reaction is carried out in the presence of a ketone, two ozonides are produced.

$$R^1CH{=}CHR^1 + O_3 \xrightarrow{R^1CO} [R^1CHO + R^1\overset{+}{C}HO{-}\overset{-}{O} + R_2CO]$$

$$\longrightarrow R^1CH(O{-}O)(O)CHR^1 + R^1CH(O{-}O)(O)CHR^2_2$$

Furthermore, if the alkene is unsymmetrical, then two carbonyl compounds and two dipolar ions are possible and their combination can lead to three different ozonides. Also, because the ozonides have cyclic structure, each one can exist in both *cis*-and transforms. Thus, six ozonides are possible from alkenes of the type $R^1R^2C = CR^3R^4$.

The ozonide is prepared by dissolving the olefinic compound in a solvent that is unaffected by ozone, *e.g.,* chloroform, carbon tetrachloride, glacial acetic acid, light petrol, etc., and a stream of ozonized oxygen is passed through. Subsequent treatment may be one of the following procedures.

(i) The ozonide is oxidised by means of silver oxide, hydrogen peroxide or per-acids, there by producing acids and/or ketones.

(ii) (a) Reduction of the ozonide with zinc-acid, H2-Raney nickel, triphenylphosphine, *etc.*, gives aldehydes and/or ketones. (b) Reduction may also be carried out with lithium aluminium hydride or sodium borohydride; the products are the corresponding alcohols of the carbonyl compounds formed in (a).

$$R^1_2C\overset{O-O}{\underset{O}{<\quad>}}CHR^2 \begin{cases} \xrightarrow{(i)} R^1_2CO + R_2CO_2H \\ \xrightarrow{(a)} R^1_2CO + R^2CHO \\ \xrightarrow{(b)} R^1_2CHOH + R^2CH_2OH \end{cases}$$

Unlike mono-di-, and tri-substituted ethylenes, tetra-alkyl-substituted ethylenes do not give normal ozonides on ozonolysis by the usual procedures; the product is a polymeric peroxide, (-CR_2-O-O). It would therefore appear that the dipolar ion, R_2CO_2, cannot combine with a ketone. However, Nebel (1968) has shown that ketones may be isolated in high yield by the addition of water to the ozonolysis products of a tetra-alkyl-substituted ethylene in an alcohol at low temperature.

$$R^1_2C{-}O{-}O + R_2OH \longrightarrow R^1_2CO(OR_2)OOH \xrightarrow{H_2O} R^1_2CO + R^2OH + H_2O_2$$

The complete process of preparing the ozonide and decomposing it (and identifying the products formed) is known as ozonolysis, and is used for determining the position of a double bond in any olefinic compound. Ozonolysis is also used as a means of preparing aldehydes, ketones and acids, and chromatography, particularly GLC and TLC, has been used to isolate the products. There is, however, an increasing tendency to use the Lemieux reagent instead of ozonolysis.

Alkenes isomerise when heated at high temperature (500-700°C), or at a lower temperature (200-300°C) in the presence of various catalysts ,*e.g.*, aluminium chloride. Isomerisation may be due (i) to the change in position of the double bond, which always tends to move towards the centre of the chain, *e.g.*, pent-1-ene isomerises to pent-2-ene:

$$CH_3CH_2HC_2CH{=}CH_2 \; AlCl_3 \rightleftharpoons CH_3CH_2CH_2\overset{+}{C}HCH_2{-}AlCl_3 \underset{+H^+}{\overset{-H^+}{\rightleftharpoons}}$$

$$CH_3CH_2HC_2CH{=}CHCH_2{-}\overset{-}{A}lCl_3 \underset{-H^+}{\overset{H^+}{\rightleftharpoons}} CH_3CH_2CH{=}CHCH_3 + AlCl_3$$

(iii) To the migration of a methyl group, *e.g.*, but-1-ene isomerises to isobutene:

$$\underset{\substack{|\\ \text{Me}}}{CH_2}\text{—}CH=CH_2 \;\; AlCl_3 \rightleftharpoons \overset{\text{Me}}{\overset{|}{CH_2}}\text{—}\overset{+}{CH}\text{—}CH_2\text{—}\overset{-}{AlCl_3}$$

$$\rightleftharpoons CH_2\text{—}\overset{\text{Me}}{\overset{|}{\overset{+}{C}}}\text{—}CH_2\text{—}\overset{-}{AlCl_3} \rightleftharpoons CH_2=CMe_2 + AlCl_3$$
$$\qquad\quad H$$

(i) and (ii) may, or may not, occur together.

Hydroboronation (hydroboration) : Alkenes react rapidly with diborane in, *e.g.*, ether at room temperature to form trialkylboranes. Terminal alkenes give primary alkynoranes which, on oxidation with alkaline hydrogen peroxide, produce primary alcohols (Brown *et al.*, 1957). Also the overall hydration of the alkene has been shown to occur by *cis*-addition.

$$RCH=CH_2 \xrightarrow{F_2H_2} (RCH_2CH_2\text{—})_3B \xrightarrow[NaOH]{H_2O_2} 3RCH_2CH_2OH + H_3BO_3$$

The anti-Markownik off's direction of addition to form the primary trialkylborane may be explained by the fact that borane is an electron-deficient molecule and so behaves as an electrophile, tending to attack at the point of highest electron density. To explain the overall *cis*-hydration, however, is not so simple. A possible mechanism, is that hydroboronation is stereospecifically *cis*-addition, and this would be explained by the reaction proceeding step wise through an intermediate cyclic T.S.

```
R   H                  R   H              R H
 \ /      H             \ /                 |
  C <---  |              C ------H          C—H
  ||      B—H    →       ||      |    →     |
  C --->  |              C ---- B—H         C—B—H
 / \      H             / \      |          |   |
H   H                  H   H     H          H   H
```

$$\xrightarrow{RCH=CH_2} (RCH_2CH_2\text{—})_2\,BH \xrightarrow{RCH=CH_2} (RCH_2CH_2\text{—})_3\,B$$

The oxidation must therefore occur by conversion of the C-B bond into C-OH without inversion. A possibility is one which involves a 1,2-shift (see p. 146)

$$H\text{—}O\text{—}O\text{—}H + OH^- \rightleftharpoons H_2O + OOH^-$$

$$R_3B\;\; \bar{O}OH \longrightarrow R_2\overset{R}{\overset{|}{\bar{B}}}\text{—}O\text{—}OH \longrightarrow R_2B\text{—}OR + OH^-$$

$$\longrightarrow R_2BOH + RO^- \xrightarrow{H_2O} R_2BOH + ROH + OH^-$$

Since R migrates with its bonding pair, the overall hydration is *cis*.

The Makownik off direction of hydration may be carried out by the oxymercuration-demercuration method.

If the trialkylborane is treated with propionic acid, the corresponding alkane is obtained by protolysis.

$$(RCH_2CH_2—)_3B \xrightarrow{H^+} 3RCH_2CH_3$$

Thus the final product is formed by reduction of an alkene by a *non-catalytic* method.

In the same way, alkynes undergo monohydroboronation and protolysis to give almost pure *cis* alkene:

$$—C{\equiv}C— \xrightarrow{B_2H_6} (—C{=}\overset{|}{C}—)_3B \xrightarrow{H^+} —CH{=}CH—$$

Addition of diborane to a non-terminal double bond is at the least hindered carbon atom, and the nature of the product depend on the amount of hindering at the double bond, *e.g.*,

$MeCH{=}CHMe$	$Me_2CH{=}CHMe$	$Me_2C{=}CMe_2$
↓	↓	↓
$(MeCH_2CHMe—)_3B$	$(MeCHCHMe—)_2BH$	$Me_2CHCMe—BH_2$

These s- and t-alkylboranes undergo isomerisation to primary alkylboranes when heated, *e.g.*,

$$(MeCH_2CHMe—)_3B \longrightarrow (MeCH_2CH_2CH_2—)_3B$$

Bis-3-methyl-2-butylborane (disiamylnorane; second example above) is particularly useful for selective hydroboronation; this borane is often written as $Si\alpha_2BH$. Thus;

$$\underset{(B_2H_6)}{\overset{20\% \uparrow \quad \uparrow 80\%}{C_6H_5CH{=}CH_2}} \qquad \underset{(Si\alpha_2BH)}{\overset{2\% \uparrow \quad \uparrow 98\%}{C_6H_5CH{=}CH_2}}$$

In this way, a compound containing more than one double bond can be converted into one particular alcohol, *e.g.*, Brown *et al.* (1961):

$$\text{(cyclohexenyl)}CH{=}CH_2 \xrightarrow{Br_2BH} \text{(cyclohexenyl)}CH_2CH_2BSia_2 \xrightarrow[NaOH]{H_2O_2} \text{(cyclohexenyl)}CH_2CH_2OH \quad \text{(v.g.)}$$

Organoboranes may also be coupled to form alkanes.

Alkenes condense with acetic anhydride in the presence of a catalyst, *e.g.*, zinc chloride, to form unsaturated ketones, *e.g.*, ethylene forms methyl vinyl ketone:

$$CH_2=CH_2+(CH_3CO)_2O \xrightarrow{ZnCl_2} CH_2=CHCOCH_3+CH_3CO_2H$$

Acid chlorides, alkyl chlorides and α-halogenated ethers also combine with alkenes in the presence of aluminium chloride, *e.g.*,

$$CH_3COCl + CH_2 = CH_2 \xrightarrow{AlCl_3} CH_3COCH_2CH_2Cl$$

$$(CH_3)CCl + CH_2 = CH_2 \xrightarrow{AlCl_3} (CH_3)_3CCH_2CH_2Cl$$

$$CH_3OCH_2Cl + CH_2 = CH_2 \xrightarrow{AlCl_3} CH_3OCH_2CH_2Cl$$

All of these are examples of the Friedel–Crafts reaction with aliphatic compounds.

In the presence of a catalyst, *e.g.*, concentrated sulphuric acid, alkenes alkylate alkanes containing a tertiary hydrogen atom. This reaction is particularly useful for preparing 2,2,4-trimethylpentane iso-octane) by treating isobutane (4 parts) with isobutane (1 part) in the presence of concentrated sulphuric acid.

$$(CH_3)_2C=CH_2+(CH_3)_3C^+ \longrightarrow (CH_3)_3C^+$$

$$(CH_3)_3C^+ \ CH_2=C(CH_3)_2 \longrightarrow (CH_3)_3CCH_2\overset{+}{C}(CH_3)_2 \xrightarrow{(CH_3)_3CH}$$

$$(CH_3)_3CCH_2CH(CH_3)_2+(CH_3)_3C^+ \quad ; \text{etc.}$$

On the other hand, various alkenes may dimerise under the influence of suitable catalysts, *e.g.*,

$$Me_2C=CH_2 \xrightarrow[(H_2SO_4)]{H^+} Me_2C^+ \ CH_2=CMe_2$$

$$Me_3CCH_2\overset{+}{C}Me_2 \xrightarrow{-H^+} Me_3CCH=CMe_2+Me_3CCH_2CMe=CH_2$$

Under different conditions, alkenes polymerise, *e.g.*, isobutene gives a polymer:

$$nC_4H_8 \xrightarrow{H^+} (C_4H_8)_n$$

When two compounds have the same empirical formula but differ in molecular weight, the more complicated compound is called a *polymer* of the simpler one. The term *polymerisation* was used originally to indicate the process that took place when a single substance—the *monomer*—gave products having the same empirical formula but different molecular weights, each of these being a multiple of that of the monomer. As the investigation of polymerisation reactions progressed, it was found that many compounds of high molecular weight, although they produced a large number of monomer molecules on suitable treatment, did not always have exactly the same empirical formula as the parent monomer. This led to a modification of the definitions of the terms polymer and polymerisation. According to Carothers (1931) ploymerisation is best defined as *intermolecular combinations that are functionally capable of proceeding indefinitely.* This definition implies that there is not limit theoretically to the size of the polymer molecule. In practice, however, the polymer ceases to grow, for various reasons. The terms polymer and polymerisation are now used mainly in connection with high molecular weight compounds, which, in addition to being called polymers, are also known as *macromolecules.*

There are two types of polymerisation, *addition polymerisation* and *condensation polymerisation.*

Addition polymerisation. Addition polymerisation occurs between molecules containing double or triple bonds; but in certain cases it can also occur between functional compounds that result from the opening of ring structures (*e.g.*, ethylene oxide). *There is no liberation of small molecules during addition polymerisation.*

A very important group of olefinic compounds that undergo addition polymerisation is of the type $CH_2 = CHY$, where Y may be H, X, CO_2R, CN, etc.,

$$nCH_2 = CHY \longrightarrow (\text{—}CH_2CHY\text{—})_n$$

There are three possible ways in which this polymerisation can occur:

(*i*) Head to tail:—CH_2CHY—CH_2CHY—

(*ii*) Head to head and tail to tail:—$CHYCH_2$—CH_2CHY—$CHYCH_2$—CH_2CHY—

(*iii*) A random arrangement involving (*i*) and (*ii*).

Experimental work indicates that (*i*) is favoured.

Most polymerisations are carried out in the presence of catalysts, and polymerisation of alkenes can be accelerated by ionic-type catalysts or radical-type catalysts. Both types of reaction consist of a number of steps which follow one another consecutively an rapidly, and take place in three principal stages:

(*i*) $H^+ \; CH_2{=}CMe_2 \longrightarrow Me{-}\overset{+}{C}Me_2$

(*ii*) $Me_3C^+ \; CH_2{=}CMe_2 \; CH_2{=}CMe_2$

$\cdots\cdots\longrightarrow CH_2C{-}(CH_2{-}CMe_2){-}CH_2{-}\overset{+}{C}Me_2$

(*iii*) $Me_3C{-}(CH_2{-}CMe){-}CH_2{-}CHe_2$

$\xrightarrow{-H^+} Me_3C{-}(CH_2{-}CMe_2){-}CH{-}CMe_2$

This polymerisation, carried out in the presence of a small amount of butadiene, produces *butyl rubber*.

The most important ionic catalysts are those introduced by Ziegler and Natta (see below).

The free-radical mechanism of catalysis. Addition polymerisation may take place by chain reactions brought about by catalysts that are known to generate free radicals. The most widely used catalysts are the organic and inorganic perioxides and the salts of the per-acids, *e.g.*, benzoyl peroxide, acetyl peroxide, hydrogen peroxide, potassium perborate, etc.

This type of mechanism is illustrated for polymerisations in the presence of organic peroxides.

(*i*) $(RCO)_2O_2 \longrightarrow 2RCO_2 \longrightarrow 2R + 2CO_2$

$R + CH_2 = CH_2 \longrightarrow R{-}CH_2{-}CH_2$

(*ii*) $R{-}CH_2{-}CH_2 + nCH_2 = CH_2 \longrightarrow R{-}(CH_2{-}CH_2)_n{-}CH_2{-}CH_2$

(*iii*) The termination reaction may take place in various ways. One obvious way is by the combination of two growing chains, *i.e.*,

$$2R{-}(CH_2{-}CH_2)_n \longrightarrow R{-}(CH_2{-}CH_2)_{2n}{-}R$$

On the other hand, termination may also occur by disproportionation:

$$2R{-}(CR_2{-}CH_2)_n{-}CH_2{-}CH_2 \longrightarrow R{-}(CH_2{-}CH_2)_n{-}CH_2CH_3 + R{-}(CH_2{-}CH_2)_n{-}CH = CH_2$$

Still another way for termination to occur is by reaction between the growing chain and another species that has been added in relatively small amount before polymerisation.

$$R{-}(CH_2CH_2)_n + YZ \longrightarrow R{-}(CH_2{-}CH_2)_n{-}Y + Z$$

$$Z + CH_2 = CH_2 \longrightarrow Z\text{—}CH_2\text{—}CH_2 \longrightarrow Z\text{—}(CH_2\text{—}CH_2)_n$$
$$Z\text{—}(CH_2\text{—}CH_2)_n + YZ \longrightarrow Z\text{—}(CH_2\text{—}CH_2)_n\text{—}Y + Z$$

Polymers of the type Z—M_n—Y, *i.e.*, those which contain a 'foreign' molecule in addition to the recurring unit, are known as *telomers*, and the process of their formation is called *telomerisation.*

In 'uncatalysed' polymerisation, *i.e.*, in the absence of foreign substances, initiation may begin by *dimerisation* of the monomer:

$$2CH_2 = CH_2 \longrightarrow CH_2\text{—}CH_2\text{—}CH_2\text{—}CH_2$$

Not only can addition polymerisation take place between molecules of one kind, but it can also take place between molecules of two kinds, when the phenomenon is known as *copolymerisation* or *interpolymerisation.*

Condensation polymerisation or polycondensation. In condensation polymerisation, bi- or poly-functional molecules condense with one another, and in doing so repeatedly eliminate a small molecule such as water, ammonia, hydrogen chloride, etc., as the reaction proceeds. This type of polymerisation takes place by a series of steps, and is discussed in various parts of the text (for example, the aldol condensation, the esterification of glycols with dibasic acids, etc.)

Polymers usually possess a certain amount of crystallinity, and their tensile strength increases with molecular weight. Also, the greater the crystallinity, the greater is the tensile strength, the lower is the solubility and the higher is the m.p.

Polymers may be classified into three groups:

(*i*) *natural, e.g.*, rubber, proteins, cellulose; (*ii*) *semi-synthetic, e.g.*, nitrocellulose, cellulose acetate; and (*iii*) *synthetic, e.g.*, nylon, bakelite, perspex.

Plastics form a group of high polymers which have a fair range of deformability and mouldability, particularly at high temperatures. In plastics the polymers formed do not all have the same molecular weight, and since the polymers are not amenable to the ordinary methods of separation, the molecular weight of a 'polymer' is the *average* molecular weight. Polymerisation is carried out with the object of building up compounds with predicted properties, and since the properties of a plastic depend on the degree of polymerisation it is necessary to stop polymerisation when the desired average molecular weight is reached.

This may be done by various means, *e.g.*, variation of the concentration of the catalyst. The average molecular weight of plastics varies from about 20,000 (*e.g.*, nylon) to several hundred thousand (*e.g.*, polyvinyl chloride, 2,50,000).

Plastics are generally tough, resistant to the action of acids and alkalis, and not very much affected over a fair range of temperature. They can be moulded to any desired shape or form.

Plastics are of two main types, *thermoplastic* and *thermosetting.* Thermoplastics are linear polymers which are soluble in many organic solvents, and which soften on heating and become rigid on cooling. The process of heat-softening, moulding and cooling can be repeated as often as desired, and hardly affects the properties of the plastics. Typical thermoplastics are cellulose acetate, nitrocellulose and vinyl polymers such as polythene, perspex, etc.

Thermosetting plastics are three-dimensional polymers which are insoluble in any kind of solvent, and which can be heat-treated only once before they set, *i.e.*, their formation, after which heating results in chemical decomposition, and hence they cannot be 'reworked'. Typical thermosetting plastics are phenol-formaldehyde, rea-formaldehyde, melamine-formaldehyde, silicones, etc.

In thermoplastics the chains are, more or less, free chemically, but are held together by vander Waals forces. It is possible, however, to link together these linear molecules (*cf*, the rungs of a ladder) and the cross-linking agent converts the thermolplastic into a thermosetting plastic, *e.g.*, in the vulcanisation of rubber the sulphur cross-links the long chains. Furthermore, such thermosetting plastics may be reconverted into thermoplastics by opening the cross-links, *e.g.*, the reclaiming of rubber. Most thermosetting plastics may be regarded as cross-linked polymers.

Those plastics which do not soften very much with rise in temperature are made soft and readily workable by the addition of certain compounds known as *plasticisers*; *e.g.*, polyvinyl chloride is extremely stiff and hard, but addition of tricresyl phosphate makes it soft and rubber-like.

Stereochemistry of polymers : If we consider the polymerisation of the monomer, $CH_2 = CYZ$, then there are two extreme stereoisomers possible : (*i*) all the Y group lie on one side of the chain, and all the Z groups on the other side: this is the *isotactic* polymer: (*ii*) the Y and

Z groups lie alternately on each side of the chain: this is the *syndiotactic* polymer.

Y Z Y Z — C — C — (chain) CH₂ CH₂ CH₂

isotactic

Y Z Z Y — C — C — (chain) CH₂ CH₂ CH₂

syndiotactic

Between these two extremes, there are those polymers in which the two groups Y and Z are arranged in a random fashion along the two sides of the chain: these are the *a tactic* polymers.

Ziegler-Natta catalysts are stereospecific: they are metal compounds, *e.g.*, $TiCl_4$, $Zr(OR)_2$, etc. together with co-catalysts such as R_3Al, $LiAlH_4$, etc., *e.g.*, a typical Zielger-Natta catalyst is the tri-ethylaluminium-titanium chloride complex.

$$Et_3Al + CH = CH_2 \longrightarrow EtCH_2CH_2AlEt_2 \xrightarrow{(CH_2=CH_2)}$$

$$Et(CH_2CH_2)_nCH_2CH_2AlEt_2 \xrightarrow{H^+} Et(CH_2CH_2)_nCH_2CH_3$$

It should be noted, however, that a stereochemically pure polymer is not usually obtained, the product being a mixture of predominantly the stereospecific product together with some atactic compound.

Some Individual Alkenes

Diradicals (biradicals) : Electrons have a spin number of ± 1/2. When all the electrons in a molecule are paired, each pair has the spins anti-parallel (+ 1/2 and – 1/2) and so cancel each other, the result being a diamagnetic molecule, *i.e.*, one which does not possess a permanent magnetic moment. Now suppose one of the paired electrons is raised by *e.g.*, electromagnetic radiation, to a higher electronic level.

In the excited molecule the spins may still be antiparallel and so cancel out. The molecule is therefore still diamagnetic. The spins may, however, be parallel (this is possible because the two electrons are in different orbitals), and the molecule will now be paramagnetic.

If S is the algebraic sum of the spins of the electrons in a molecule, then M, the *multiplicity,* is given by the expression

$$M = 2S + 1$$

In covalent pairs,

$$S = (+ 1/2) + (-1/2) = 0$$

Therefore $M = 1$, and the molecule is said to be in a *singlet* state. If the molecule has *one odd electron,* $S = ½$ and therefore $M = 2$; the molecule is now in a *doublet* state. If the molecule ha two electrons in different orbitals then, if the spins are antiparallel, $S = 0$, $M = 1$, and the molecule is in a single state.

If the spins are parallel, $S = (1/2) + (+1/2) = 1$, $M = 3$ and the molecule is in a *triplet* state; in this state the two electrons interacts strongly with each other.

There is also the case where the interaction (coupling) is very weak or absent, *i.e., the unpaired electrons act independently of each other; in the case, the molecule is a diradical* (one might call it a 'double doublet'). A molecule in the triplet state can behave as a diradical in chemical reactions.

Methylene(carbene), CH_2 : This is the first member of the alkenes, but it has only a very short life, and is a 'bivalent' carbon compound. It is formed by the photolysis (photochemical decomposition) or pyrolysis of diazomethane or keten:

$$CH_2N_2 \longrightarrow CH_2 + N_2$$
$$CH_2{=}C{=}O \longrightarrow CH_2 + CO$$

Methylene undergoes two types of reaction, insertion and addition. Insertion reactions occur mainly in the C—H bond, but can also occur in O-H and C-Cl bonds *e.g.,* (Strachan *et al.,* 1954, Bradley *et al.,* 1961):

$$CH_3CHOHCH_3 \xrightarrow{CH_2} (CH_3)_3COH + CH_3CH_2CHOHCH_3 + (CH_3)_2CHOCH_3$$

$$CH_3CH_2CH_2Cl \xrightarrow{CH_2} CH_3(CH_2)_2CH_2Cl + CH_3CH_2CHClCH_3$$

The usual mechanism for insertion is believed to be via a cyclic transition state *e.g.,*

$$H{-}CH_2{-}CH\langle + CH_2 \longrightarrow H \cdots CH_2{-}CH\langle \ (\cdots CH_2) \longrightarrow CH_3{-}CH_2CH\langle$$

Methylene adds across double bonds to form cyclopropanes; at the same time, insertion reactions also occur. Skell *et al.,*(1956, 1959) showed the addition of methylene (from diazomethane) to *cis-* and *trans-* but-2-ene is a *cis*-addition. Thus the addition is stereospecific, *i.e.,* each geometrical isomer forms one product, and the configurations of the two products are different.

Me Me Me Me
C=C $\xrightarrow{CH_2}$ C——C
H H H CH_2 H
cis *cis*

Me H Me Me
C=C $\xrightarrow{CH_2}$ C——C
H Me H CH_2 H
trans *trans*

Anet *et al.*,(1960), however, showed that this stereospecific addition is lost when the reaction is carried out in the presence of an inert gas (nitorgen), *i.e.*, each substrate (reactant) now gives a mixture of the *cis*- and *trans*-products. On the other hand, Duncan *et al.* (1962) showed that methylene formed by the photolysis of keten added to the above substrates in a non-stereo-specific manner.

To explain these results, let us first consider the problem of the *relationship between reactivity and selecticity.* A general (empirical) principle is that the more reactive the reagent is, the less selective it is in its reactions, *e.g.,* in the chlorination of alkanes, the rate of hydrogen substitution is t > s > primary . If the temperature is raised above 300^0 C, chlorine becomes more reactive and the substitution now becomes less selective, the rates of substitution being the same for primary-s- and t-hydrogen. A possible basis for this *reacitive-selectivity* principle is that the more reactive the reagent is, the more likely it will react at every collision, *i.e.*, there would be less discrimination between the various positions. The less reactive the reagent, the more will the electronics distribution in the substrate play a part, thereby resulting in more discrimination at the various positions.

The above non-stereospecific additions are an indication that in an inert gas, methylene becomes less reactive, it also follows that methylene generated from keten is less reactive than that from diazomethane. The latter conclusion had already been reached by Frey (1958), who found that methylene from the photolysis of diazomethane was less selective in its insertion in primary and secondary C—H bonds than was methylene from keten. It therefore appears that there are two forms of methylene. Herzberg *et al.* (1961) showed, from spectroscopic evidence, that methylene (from diazomethane) was initially formed in the singlet state (this is the excited state) and some then rapidly changes to the triplet state (this is the ground state).

The electronic distribution in methylene has been the subject of much discussion. It appears that the general feeling now is that in the singlet state, the carbon atom is sp^2 hybridised with two electrons in each of these three orbitals, whereas in the triplet state, the carbon is sp hybridised with two electrons in each of these orbitals, and the two p-orbitals each occupied by one electron. Thus the singlet state would be bent and the triplet state would be linear. The singlet, with its vacant orbital, can be expected to be eletrophilic (*cf.* carbonium ions). On the other hand, the triplet is a diradical type, and can be expected to behave as a free radical.

The details of the mechanism for the addition reactions is still in a state of flux, but a widely held theory is that of Skell *et al.*, (1956). According to this theory, the stereospecificity is due to the addition of singlet methylene in one step:

$$>C{=}C< + CH_2 \longrightarrow \underset{CH_2}{>C{=}C<} \longrightarrow \underset{CH_2}{>C{-}C<}$$

The non-stereospecific addition involves *triplet* methylene. In this case, a cyclic transition state is not formed, the addition occuring step wise as a radical addition. This is represented as follows, the arrows indicating the direction of spin of each unpaired electron; and the assumption is made that rotation about the single bond is more rapid than spin inversion:

$$RHC{=}CHR + CH_2 \longrightarrow RHC^{\downarrow}{-}CHR(CH_2^{\downarrow}) \longrightarrow RHC^{\uparrow}{-}CHR(CH_2^{\downarrow}) \longrightarrow \underset{CH_2}{RHC{-}CHR}$$

$$\rightleftharpoons$$

$$\longrightarrow \underset{CH_2}{RHC{-}CHR} \quad HHC^{\downarrow}{-}CHR(CH_2^{\downarrow}) \longrightarrow HHC^{\uparrow}{-}CHR(CH_2^{\downarrow}) \longrightarrow \underset{CH_2}{HHC{-}CHR}$$

Ethylene may be prepared by most of the general methods of preparation, but the most convenient laboratory method is to heat ethanol with excess of concentrated sulphuric acid. It is obtained in huge quantities as a by-product in the cracking of crude oil and of ethane and propane.

Ethylene is a colourless gas, b.p. –105°C, sparingly soluble in water. It burns with a smoky luminuous flame. When ethylene is heated with chlorine at 350–450°C, vinyl chloride is obtained:

$$CH_2 = CH_2 + Cl_2 \longrightarrow CH_2 = CHCl + HCl$$

Ethylene may be oxidised to ethylene oxide, and polymerises under high pressure and high temperature to form *polyethylene* or *polythene:*

$$nCH_2 = CH_2 \longrightarrow -(-CH_2-CH_2-)n-$$

This polymerisation is catalysed by traces of oxygen (which produces the free radicals). Polythene is very resistant to acids, bases, and most or the usual organic solvents.

Polythene is also manufactured by the Ziegler process. Ethylene is passed, under pressure, into a hydrocarbon solvent containing a suspension of, *e.g.*, $Et_3Al + TiCl_4$ as catalyst at about 70°C.

Ethylene is used for ripening fruit. Unripe fruit may be transported easily without damage, and ripens on exposure to ethylene gas for a few days, the product being apparently indistinguishable from the natural ripened fruit. It appears that all fruits give off ethylene (Burg *eg al.*, 1962). Ethylene is also used as an anaesthetic, in the manufacture of mustard gas and plastics (polythene, polystyrene), and in the preparation of various solvents such as glycol, dioxan, cellosolves, etc.

Structure or ethylene : The molecular formula of ethylene is C_2H_4. Two carbon atoms have the power to combine with six univalent atoms or groups, as in ethane, neopentane, *etc.* There are only four univalent hydrogen atoms present in ethylene; therefore ethylene is said to be unsaturated, and should be capable of adding on two univalent atoms or groups, and this, as we have seen above, is observed in practice. Thus the structure of ethylene must be such as to be capable of undergoing addition reactions. Assuming carbon to be quadrivalent and hydrogen univalent, three structures are possible for ethylene;

$$\underset{(I)}{CH_3-CH_2} \qquad \underset{(II)}{CH_2 = CH_2} \qquad \underset{(III)}{\cdot CH_2-CH_2\cdot}$$

Two isomeric compounds of molecular formula $C_2H_4Cl_2$ are possible; CH_3CHCl_2 and CH_2ClCH_2Cl. Both isomers are known, one (ethylene dichloride) being formed by the direct combination between ethylene and chlorine, and the other (ethylidene dichloride) by the action of phosphorus pentachloride on acetaldehyde. The structure of ethylidene dichloride is CH_3CHCl_2 ; hence the structure of ethylene dichloride is CH_2ClCH_2Cl. If (I) were the structure of ethylene, then the addition of chlorine should give ethylidene dichloride, and not ethylene dichloride. We may, therefore, reject structure (I). Furthermore, since (I) is unsymmetrical it would have a fairly large dipole moment; actually ethylene has a zero dipole moment.

Structure (II) represents ethylene as possessing a *double bond.* Such a bond would prevent free rotation and would therefore explain geometrical isomerism. Hence (II) was acceptable to the classical chemists. The modern theory of a double bond is that it consists or one σ-and one π-bond, or two 'bent' bonds. Structure (III) represents ethylene as a free diradical, but since ethylene does not exhibit the usual properties of a diradical, we must reject (III).

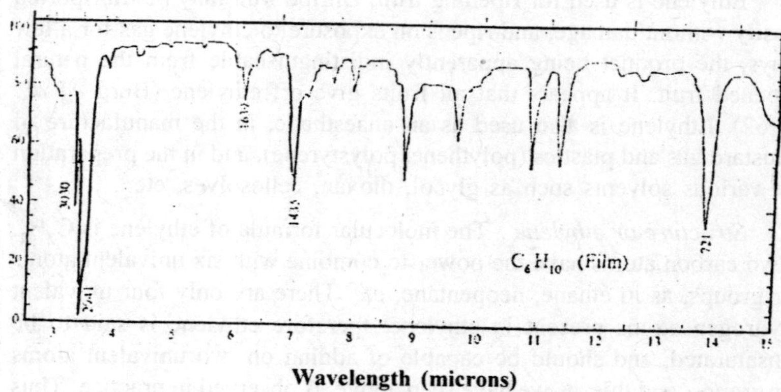

Fig. 2.1

1639 cm^{-1}; C=C (str.). 3030cm^{-1}; = C—H(str.). 721 cm^{-1}; =C—H (def.); *cis* isomer, 2941 cm^{-1} C–H (str.) in (Me and/or) CH_2. 1435 cm^{-1}; C—H (def.) in (Me and/or) CH_2.

The presence of a double (or triple bond) in an organic compound may be found readily by means of bromine water, bromine in chloroform solution, or dilute alkaline permanganate. If the compound under investigation is unsaturated, then the above reagents are decolorised. Perbenzoic acid or monoperphthalic can be used to detect the presence of a double bond, and also to estimate the number of double bonds (see also iodine value). The presence of an isolated double bond is readily detected by its infrared absorption spectrum (C = C stretch), 1680-1620 cm^{-1} (v), *cis*-and *trans*-Isomers or the type RCH=CHR may be distinguished by the C–H deformation frequencies; *cis,* 630-665 cm^{-1} (s); trans, 970-960 cm^{-1}(s). Also, there are several regions of absorption due to the =C–H stretch. The actual positions depend on the nature of the alkene; mono-, s-di-(both *cis* and *trans*) and trisubstituted ethylenes all show 2 band at 3040-3010 cm^{-1} (m), and as-disubstituted ethylenes shown a band at 3095-3075 cm^{-1} (m) [this bond is also shown by monosubstituted ethylenes].

NMR spectra of a alkenes : Since the chemical shift of an olefinic proton (t = 4.3-5.3p.p.m.) is very different from that of a methyl, methylene, proton (t = 7.0-9.1) in hydrocarbons, and from aromatic (and heterocyclic) protons (t = 1.0-3.0), it is relatively simple to detect the presence of a double bond in a compound (olefinic) containing the = CH group. Fig. 2.2 is the NMR spectrum of cyclohexene (in CCl_4 at 60 MHz).

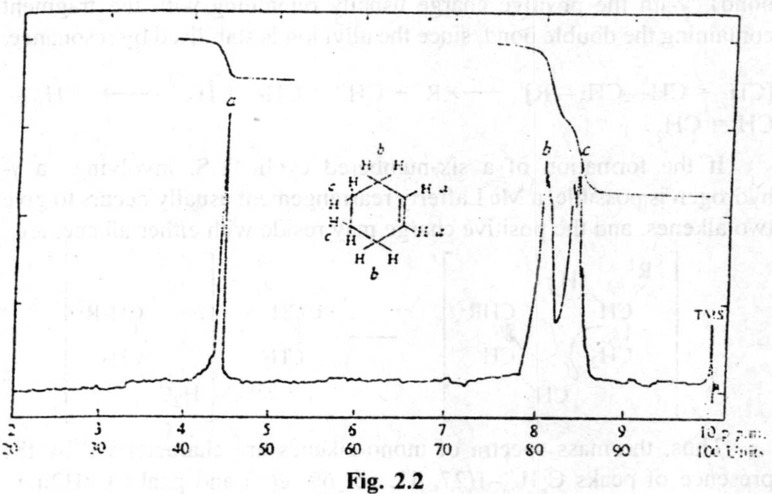

Fig. 2.2

It has been pointed out that the spin-spin coupling constant depends on the number of covalent bonds between the two interacting protons and on their geometry. The former factor is the same for a pair of *cis*-*trans* 1,2-disubstituted ethylenes.

The latter factor, however, is different, and this produces different *J* values and so offers a means of distinguishing between *cis* and *trans* isomers.

H H C=C	H C=C H	H C=C H
cis	*trans*	*geminal*
J 7-14 Hz	12-19 Hz	0-3 Hz

J-values for other groupings involving a double bond are, *e.g.* (note the long range coupling):

CH C=C H	CH C=C H	H C=C H
vicinal	allylic	homo-allylic
J 4-10 Hz	(*cis or trans*) 0-2 Hz	0-2 Hz

The molecular ion of mono-alkenes is usually present and tends to undergo allylic cleavage (*i.e.,* at the β-bond with respect to the double bond). With the positive charge usually remaining with the fragment containing the double bond, since the allyl ion is stabilised by resonance.

$[CH_2 = CH—CH_2—R] \longrightarrow R + CH2 = CH—CH_2^+ \longleftrightarrow \overset{+}{C}H_2—CH = CH_2$

If the formation of a six-numbered cyclic T.S. involving a γ-hydrogen is possible, a Mc Lafferty rearrangement usually occurs to give two alkenes, and the positive charge may reside with either alkene, *e.g.,*

$$\left[R^1CH(H)—CH_2—CH_2—CH{=}CHR^2 \right]^+ \longrightarrow R^1CH{=}CH_2 + \left[CH_2R^2—CH_2—\dot{C}H_2 \right]^+$$

Thus, the mass spectra of mono-alkenes are characterised by the presence of peaks $C_nH_{2n}-1$(27, 41, 55, 69, *etc.*) and peaks CnH2n,

A difficulty encountered with alkenes is that, because of the ready migration of the double bond, fragmentation of isomeric alkenes are often similar.

Propene (propylene) : May be prepared by heating *propanol* or *isopropanol* with sulphuric acid (mechanism as for ethylene from ethanol):

$$CH_3CH_2CH_2OH \xrightarrow{-H2O} CH_3CH{=}CH_2$$

$$CH_3CH(OH)CH_3 \xrightarrow{-H2O} CH_3CH{=}CH_2$$

It may also be prepared by heating propyl iodide with ethanolic potassium hydroxide:

$$CH_3CH_2CH_2I{+}KOH \xrightarrow{\text{ethanol}} CH_3CH{=}CH_2 + KI + H_2O$$

Propene is obtained commercially in huge quantities as a by-product in the cracking of petroleum.

It is a colourless gas, b.p. –48°C, insoluble in water but fairly soluble in ethanol is used industrially for the preparation of isopropanol, glycerol, polypropylene (a plastic), *etc.*

Butenes : There are three isomeric butenes, and all are gases: $CH_3CH_2CH = CH_2$, but -l-ene (b.p.-6.1°C); $CH_3CH{=}CHCH_3$, but-2-ene (b.p. 1°C)

All the butenes are obtained from craked petroleum. The 1- and 2-butens are used for the preparation of s-butanol (q.r), and isobutene for t-butanol (q.v). But-2-ene differs from its isomers in that it exhibits geometrical isomerism.

Isobutene differs from its isomers in that it reacts with chlorine *at room temperature* to give mainly *substitution* products, substitutions occuring in the allyl position. Thus 3-chloro-2-methylprop-l-ene is the main product, and it accompained by a small amount of the addition product 1.2-dichloro-2-methylpropane:

$$(CH_3)_2C{=}CH_2 \xrightarrow{Cl_2} (ClCH_2)(CH_3)C{=}CH_2 + (CH_3)_2CClCH_2Cl$$

Mechanistic studies have shown that the substitution product is not obtained by *direct* substitution Reeve (1952), using isobulate labelled at position 1 (^{14}C) did not obtain labelled formaldehyde on ozonolysis of the monochloro isobutene.

Since the dichloride is stable under the conditions of the experiment, a mechanism which is in accord with these facts is:

$$Me{-}C(CH_3){=}\dot{C}H_2 \quad Cl{-}Cl \longrightarrow Me{-}\overset{+}{C}(CH_2{-}H)\,{-}\dot{C}H_2Cl \quad Cl^- \longrightarrow HCl + Me{-}C(={CH_2}){-}\dot{C}H_2Cl$$

UNSATURATED COMPOUNDS WITH TWO OR MORE DOUBLE BONDS

When the compound contains two double bonds, it is known as a diolefin or alkadiene, and has the general formula $C_\pi H_{2\pi-2}$; when there are three double bonds present, the compound is known as a triolefin or alkatriene, and has the general formula $C_\pi H_{2\pi-4}$; etc.

Hydrocarbons with isolated double bonds contain the arrangement $>C{=}CH(CH_2)_pCH{=}C<$ where $p > 0$. One of the simplest compounds of the type is diallyl or hexa-1, 5-diese which may be prepared by the Wurtz reaction with allyl chloride:

$$2CH_2 = CHCH_2Cl + Mg \xrightarrow{\text{ether}} CH_2 = CH(CH_2)_2C = CH_2 + MgCl_2\ (56\text{-}65\%)$$

Hexa-1.5-diene is a liquid, b.p. 59.6°C. Alkadienes with *isolated* double bonds react like the alkenes each olefinic bond acting independently of the other.

Hydrocarbons with comulated double bonds contain the arrangement $>C{=}C{=}C<$. The simplest compound of this type is allene or propadiene, which may be prepared by heating 1.2.3-tribromopropane with solid potassium hydroxide, and the treating the resulting 2.3-dibromopropene with zinc dust in methanol.

$$BrCH_2CHBrCH_2Br \xrightarrow{\text{KOH}} BrCH_2CBr{=}CH_2 \xrightarrow{\text{ZaCH}_3\text{OH}} CH_2 = C = CH_2$$

Allene is a gas, b.p. – 32 C. With bromine it forms 1.2.2.3-tetrabromopropane; with sulphuric and it forms acetone and when treated with sodium in ether, the sodium derivative of propyne is produced.

$$\text{(a)}\ CH_2 = C = CH_2 \xrightarrow{H^+} CH_3\overset{+}{C} = CH_2 \xrightarrow{H_2O}$$

$$CH_3C(OH_2) = CH_2 \xrightarrow{-H^+} CH_3C(OH) = CH_2 \longrightarrow CH_3COHC_3$$

$$\text{(b)}\ CH_2 = C = CH_2 \xrightarrow[(-H^+)]{Na} CH_2 = C = \dot{C}H_2 \;—\; C \equiv CH$$

$$\xrightarrow{H^+} CH_3C \equiv CH \xrightarrow[(-H^+)]{Na} CH_3C \equiv C^-Na^+$$

Alkenes may be prepared from allenes by reversal of the above are arrangement, *e.g.*,

$$CH \equiv CCH_2CO_2H \xrightarrow{aq\ K_2CO_3} CH_2 = C = CHCO_2H$$

A novel way of preparing allenes is by treated the cyclopropane derivative formed from an alkene and bromoform and alkali with magnesium ether (Docering et al., 1958). Dibromomethylene is an intermediate.

$$R_2C{=}CR_2 + :CBr_2 \longrightarrow R_2C\underset{CB_2}{\overset{}{-\!\!-}}CR_2 \xrightarrow{Mg} R_2C{=}C{=}CR_2$$

An extended alkene type of linkage gives the *cumulene* system, the simples member of which is butariene, and this has been prepared by debrominating 1,4-dibromobut-2-yne with zinc) Schubert et al., 1952, 1954).

$$BrCH_2C \equiv CCH_2\ Br \xrightarrow{Zn} CH_2 = C = C = CH_2$$

Alkenes are very important from the stereochemical point of view (see p. 498)

Hydrocarbons with conjugated double bonds contain single and double bonds arranged **, *i.e.*, they contain the arrangement —CH = CH—CH = CH—. The simplest member of this group of compounds is buta-1.3-diene, which may be prepared by passing cyclohexene over a heated nichrome wire (an alloy of nickel ,chromium and iron):

$$\text{cyclohexene } (CH_2\text{–}CH_2\text{–}CH_2\text{–}CH_2\text{–}CH{=}CH) \longrightarrow CH_2{=}CH\text{–}CH{=}CH_2 + CH_2{=}CH_2 \quad (65\text{-}75\%)$$

Butadiene is prepared technically in various ways, *e.g.* (i) By dehydrogenating n-butane (from nature gas or petroleum gas) over a heated chromum oxide-alumina catalyst. (ii) By dehydrogenating but-1-ene and but-2-ene (from cracked petroleum) over a heated mixture of nickel and calcium phosphates (plus a little chromic oxide) as catalyst.

Butadiene is a gas, h.p. –4.4° C. Under the influence of sodium as catalyst, butadiene readily polymerises to a product which has been used as a rubber substitute known as *buna* (but adiene-Na).

A very important alkadiene is isoprene or 2-methyl but-1,3-duiene, $CH_2 = CmeCH = CH_2$, which may be obtained, poor yield, by the slow distillation of rubber.

Isoprene is manufactured in various ways, *e.g.* (i) By passing ispoentane or isopentene over heated Cr_2O_3—Al_2O_3. (ii) Propene is dimerised by heating at 200°C under pressure in they presence of this isopropyl aluminium as catalyst, the dimer, 2-methylpen-1-ene, then isomerised to 2-methylpent-2-ene by heating to about 300°C in the presence of a catalyst, and this compound, mixed with steam and small amount of hydrogen bromide, is heated to about 800° C.

$$2CH_2{=}CHCH_3 \longrightarrow CH_2{=}C(CH_3)CH_2CH_2CH_3 \longrightarrow$$

$$CH_3C(CH_3){=}CHCH_2CH_3 \longrightarrow CH_2{=}C(CH_3)CH{=}CH_2 + CH_4$$

Isoprene is a liquid, b.p. 35°C, and when heated with sodium at 60° C it polymerises to a substance resembling natural rubber.

$$CH_2{=}C(CH_3){-}CH{=}CH_2 + Na \longrightarrow \overset{+}{Na}\overset{-}{C}H_2{-}C(CH_3){=}CH{-}CH_2$$

$$\downarrow CH_2{=}C(CH_3)CH{-}CH_2$$

$$\overset{+}{Na}\overset{-}{C}H_2{-}C(CH_3){=}CH{-}CH_2{-}CH_2{-}C(CH_3){=}CH{-}CH_2,$$

Compounds containing conjugated double bonds have physical and chemical properties but are not usually shown by compounds containing isolated or cumulated double bonds, *e.g.*, they undergo abnormal addition reactions, readily polymerise, and undergo the diels-Alder reaction. Butadiene also exhibits a special case of the diels-Alder reaction in that it dimerises to 4-vinylcyclohexene, etc.

$$\begin{matrix} CH_2 \\ /\!\!/ \\ CH \\ | \\ CH \\ \backslash \\ CH_2 \end{matrix} + \begin{matrix} CH{=}CH_2 \\ / \\ CH \\ \| \\ CH \end{matrix} \longrightarrow \begin{matrix} CH_2 \\ / \quad \backslash \\ CH \quad CH{-}CH{=}CH_2 \\ | \quad\quad \| \\ CH \quad CH \\ \backslash \quad / \\ CH_2 \end{matrix}$$

This compound regenerates but adenine heating another typical reaction of conjugated dienes is their combination with sulphur dioxide to form a cyclic sulphone, *e.g.*

$$CH_3{=}CH{-}CH{=}CH_2{-}SO_2 \longrightarrow \begin{matrix} CH{-}CH_2 \\ \| \quad\quad \backslash \\ \quad\quad\quad SO_2 \\ | \quad\quad / \\ CH{-}CH_2 \end{matrix}$$

sulpholene

Although an isolated double bond is not usually reduced by dissolving metals, conjugated systems are *e.g.*, isoprene and sodium in liquid ammonia:

$$CH_2{=}CMeCH{=}CH_2 + 2e \xrightarrow{2Na} [CH_2{-}CMe{=}CH{-}CH_2] \xrightarrow{2NH_3} CH_3CMe{=}CHCH_3 + 2NH_2^-$$

Conjugated systems also shown optical exaltation, and their absorption maximum in the inferred (C = C stretch) is shifted from 1680 to 1620 cm^{-1} for an isolated double bond to 1660-1580 cm^{-1}.

Thiele's Theory of Partial Valencies : Conjugated compounds undergo abnormal addition reactions, *e.g.*, when but adiene is treated with bromine (one molecule), two dibromo-derivatives are obtained, the 'expected' 3, 4-dibromobut-1-ene (1, 2-addition) and 'unexpected' 1,4-dibrom but -2-ene (1,4-addition).

$$CH_2{=}CH{-}CH{=}CH_2 \xrightarrow{Br_2} BrCH_2CHBrCH{=}CH_2 + BrCH_2CH{=}CHCH_2Br$$

It has been found that 1, 2-and 1, 4-additions usually take place together, and the relative amount of each generally depends on the nature of the addendum and the conditions of the experiment, *e.g.*, type of solvent temperature.

Thiele (1899) suggested his theory of *partial valencies* to account for 1,4-addition. According to Thiele, a single bond is sufficient to hold two carbon atoms together, and the two valencies of the double bond are not used completely to link the two carbon atoms, but only one valency and part of the other, leaving a surplus on each carbon atom. Thiele called this surplus valency the *residual* or *partial* valency, and if we represent it by a broken line, the formula of butadiene (and similarly for any other conjugated compound) may be written $CH_2{=}CH{-}CH{=}CH_2$. Thiele thought that the two middle partial valencies mutually satisfied each other rather than remain free. Thus the actual state of butadieneis

$$CH_2{=}CH{-}CH{=}CH_2 \quad \text{or} \quad CH_2{=}CH{-}CH{=}CH_2$$

The ends of this molecule are therefore the most active parts, and so addition of, *e.g.*, bromine will occur at these ends, first by attachment through the partial valencies, and then by each bromine atom acquiring a full valency, causing the two middle carbon atoms to utilities completely the two valencies left:

$$CH_2{-}CH{-}CH{-}CH_2 + Br \longrightarrow \underset{Br}{CH_2}{-}\overset{}{CH}{-}CH{-}\underset{Br}{CH_2} + Br$$

$$\longrightarrow \underset{Br}{CH_2}{-}CH{=}CH{-}\underset{Br}{CH_2}$$

Thiele's theory explains 1.4-additions well that it does not account at all for 1, 2-addition!

One explanation for 1, 2-and 1, 4-additions is obtained from a consideration of the stabilites of the various carbonium ion intermediates (and assumes electrophilic addition as for mono-alkenes)

$$CH_2{=}CH{-}CH{=}CH_2 \xrightarrow{Cl_2} \underset{(A)}{CH_2{=}CH{-}\overset{+}{C}H{-}CH_2Cl}$$

$$\text{or} \quad \underset{(B)}{CH_2 = CH{-}CHCl{-}\overset{+}{C}H_2}$$

(A) is secondary and (B) is a primary carbonium ion, and since (A) is more stable than (B), the former is preferred. (A) is also a resonance hybrid of (A) and (A′) and may be represented (C). This extended conjugation (of the allyl group) is absent in (B), and consequently (A) is further stabilised with respect to (B) and hence (A) is formed *i.e.*, the first step involves addition

$$\underset{(A)}{CH_2{=}\overset{+}{C}H{-}CH{-}CH_2Cl} \longleftrightarrow \underset{(B)}{\overset{+}{C}H_2{-}CH{=}CH{-}\overset{\delta+}{C}H_2Cl};$$

$$\underset{(C)}{{-}CH{=}\overset{\delta+}{C}H{-}CH_3Cl}$$

at the *terminal* carbon atom. The single representation of the resonance hybrid. (C), shows the presence of two positive centres. An ion such as this, *i.e.*, a cation with two positively charged centres, is called an *ambident cation*. Because of these two positive centres, addition of the chloride ion in the second step can occur at either centre to give 1, 2- and 1, 4-dichlorides:

$$\overset{+}{C}H_2{=}CH{=}\overset{+}{C}H{-}CH_2Cl + Cl^-$$

$$\longrightarrow CH_2{=}CHCHClCH_2Cl + ClCH_2{=}CHCH_2Cl$$

However, (A) will contribute more than (A′) to (C), *i.e.*, the resonance hybrid will carry more positive charge on the 'inner' carbon atom than on the terminal one. Hence, 1-2- addition will be expected to predominate over 1,4- i.e, the 1, 2-dichloride is the kinetically controlled product (provided the two products are not intervon vertible under the conditions of the experiment). This has been shown to be so in practice (see also below).

An alternative explanation for the course of addition of chlorine to butadiene proposes that attack occurs at a terminal carbon atom because this becomes electron-rich through the + E effect of the vary group:

$$CH_2{=}CH{-}CH{=}CH_2 \quad Cl{-}Cl \rightarrow CH_2{=}CH{-}\overset{+}{C}H{-}CH_2Cl + Cl^-$$

The argument to explain the formation of both 1, 2- and 1, 4-dichlorides is the same as above, but differs in that it does not predict which product predominates. This problem solved by consideration of the relative stabilites of the products. If the conditions permit interconversion, then the dynamically controlled product will predominate. If the conditions do not premises interconversion, them the kinetically controlled product will perdominate. Consider the following

$$-C{=}C{-}C{-}X$$

In this type of structure, halogen hyperconjugation is possible, and this makes the molecule more stable than one in which such hyper conjugation is not possible. Inspection of the two butadiense dichlorides shows that the 1, 2-product has one chlorine-hyperconjugated system where as the 1, 4-product has two. Thus, the latter isomer stable and so should be the thermodynamically controlled product if the two forms are interconvertible and consequently will be the predominant product under these conditions. These predictions have been observed experimentally, *e.g.* Muskat et.al. (1930) treated butadiene with chlorine under conditions where the two isomers were not interconvertible, an obtained about 60 percent of the 1, 2- and 40 percent of the 1, 4-product. This implies that the compound is the kinetically controlled product. Pudovic (1949) heated each isomer at 200° C and obtained the same equilibrium mixture containing about 30 percent 1,2- and 70 percent 1, 4-. Thus the, 4-isomer is the thermodynamically controlled product, and predominates under conditions interconvertibility.

The addition of the halogen acid to butadiene also produces two products, the 1, 2-and the 1,4-. Furthermore, since the proton adds on first, this adds always to the *terminal* carbon, the halogen then adding to position 2 or 4. Hydrogenbromide adds to butadiene in the absence of oxygen form a mixture of 3-bromobut-1-ene (80 percent) and

1-bromobut-2-ene. In the presence of or peroxides, these two products are again formed, but now the yields are reversed (Kharas: et al. 1936; cf. the peroxide effect,).

In the foregoing account of the reactions of but adiene, we have assumed that the molecule has the structure $CH_2 = CH—CH = CH_2$. There are however, alternative electronic structures which are *charged*. Hence butadiene is a resonance hybrid of a number of resonating structures:

$$CH_2 = CH—CH = CH_2 \longleftrightarrow CH_2 = CH—\overset{+}{C}H—\overset{-}{\ddot{C}}H \text{ « } \overset{+}{C}H_2—CH = CH—\ddot{C}H_2\text{, etc.}$$

There is still, however, another contributing structure of butadiene, viz (1). The two electrons have antiparallel spins, since the number of unpaired electrons in each resonating structure must be the same (p. 35). Thus these paired electrons would, in the ordinary way, from a covalent bond. The distance between them, however, is too great for them for forman *effective* bond. Consequently this bond is referred to as a *formal bond*, and may be represented by a dotted line (II). Because of the 'long bond' in (II), this will make very little contribution to the resonance hybrid. Since the charged structures have, apart from the charges, fewer covalent bonds than (III), they will contribute relatively little. Thus, the most important contribution is made by (III) and hence the resonance energy can be expected to be small. Actually, calculation has shown it to be about 14.64 kJ mol^{-1}. From this it follows that butadiene is best represented

$$\underset{(I)}{CH_2—CH{=}CH—\dot{C}H_2} \qquad \underset{(II)}{CH_2—CH{=}CH—CH_2} \qquad \underset{(III)}{CH_2—CH{=}\dot{C}H—CH_2}$$

as a resonance hybrid whose structure is very close to that of the classicalone. This means that the two terminal 'double' bonds have a very small amount of single bond character and that the two terminal 'double' bonds have a very small amount of single bond character and that the central 'single' bond has a very small amount of double bond character (cf. Thiele's formula).

So far we have considered the structure of conjugated compounds from V.B. theory. When we consider their structure from M.O. theory we get a different picture. Each carbon atominbutadiene had the trigonal arrangement, and Fig. 2.3 (a) shows the p_z electrons associated with each carbon atom. If the molecule is planner p_z electron of C_2 overlaps that of C_1 as much as it does that of C_3. etc. Therefore, all four p_z orbitals can be treated as foramina *M. O covering all four carbon atoms (b)*. In this conditions, a pair of electrons are no longer mainly confined to the region between two nuclei, *i.e.*, the bond formed is no longer a localised bond. The bonds produced are therefore called delocalised bonds. The wave equation for butadiene may therefore be written as : $\psi = c_1\psi_1 + c_2\psi_2 + c_3\psi_3 + c_4\psi_4$.

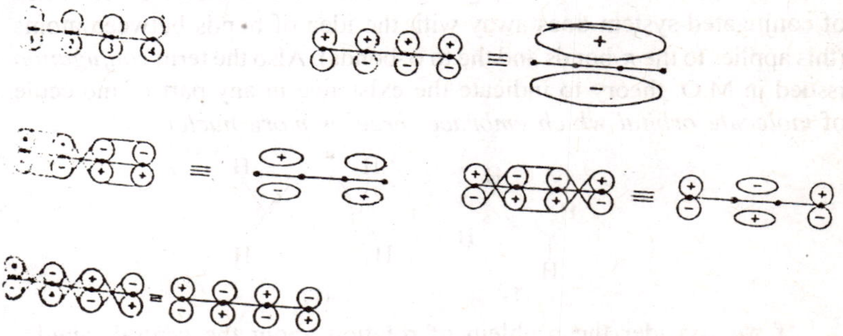

Fig. 2.3

where y_1 to y2 represent the wave functions of the four pz orbitals and c1 to are their respecting coefficients which indicate the individual contributions of the orbitals to the molecular orbital. The equation has four solutions, the values of the coefficients for each M.O. being : (b) 0-37,0-60, 0-33 (c) 0.60, 0.37,–0.37, – 0.37,– 0.37, – 0.60 (d) 0.60,– 0.37, –0.37, 0.60; (e) 0.37, –0.60, 0.60, –0.37 (b) to (e) correspond to Figs 3b to 3e; the more conventional diagrams are also given. In these diagrams the values of the coefficients have been ignored; only their signs are used, and it should be remembered that change in sign means that there is a node between the adjacent pair of atoms (this is indicated where the lines cross), as the number of nodes in anorbital increases, so does the energy associated with that orbital. Furthermore, according to the Pauli excluse principle, no more than *two* electrons can occupy the same M.O. Therefore in the ground state butadiene, two of the π-electrons will occupy the M.O. in (b), and the other two the M O. with the higher energy level, *i.e.*, (c) In any excited state of butadiene, electrons will occupy orbitals.

In delocalised bonds. The electrons have created freedom of movement than in localised bonds. Then the total energy of the system is lowered, *i.e.*, *delocalisation energy* (Coulson), but it is also usual to be it the response energy. It should here be noted that *delocalisation of bonds* is in M.O. theory the *resonance* is in V.B. theory. Furthermore, since declocaliation of electrons stabilises a system, the localisation of electrons will make a system less stable (see, *e.g.* acids and bases and aromes substitution.

It can be seen from the foregoing discussion that the M.O. treatment of conjugated system does away with the idea of bonds between atoms (this applies to the π-bonds and the to σ-bonds). Also the term *conjugation* issued in M.O. theory to indicate the existence in any part of molecule of *molecule orbital which embrace three or more nuclei.*

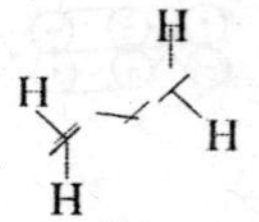

(T) 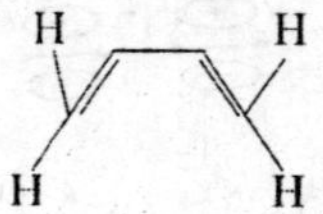 (C)

If we consider the problem of rotation about the central 'single' bond, then there are only two conformations in which all four *p*-orbitals are parallel (and therefore give maximum overlap). (T) is the *s-trans (stransoid)* form and (C) the *s*-cis (cisoid); *s* refers to the fact that there is geometrical isomers with respect to a *single* bond. According to Sheppard *et al.* (1962), butadiene exists only in the *trans* form. This is the more stable conformation owing to greater stearic repulsion (between the two H) in than in (T).

The relationship between the observed bond-length and the value expected on the assumption that a 'pure' single or double bond has been put on a quantitative basis. In the V.B. method the *double character* of a bond may be calculated from a knowledge of the observed bond length, the values of single bondinethane, and the double bond in ethylene being taken as standard lengths for 'pure' sign and double bonds, respectively.

In the M.O. method, the character of a bond is defined by its *fractional bond order*, where the best orders of 1, 2 and 3 are given to the bonds in ethane, ethylene and acetylene, respectively. Since also method of calculation is different from that of the V. B. method, the numerical values obtained by two methods are different. Even so, these values always correspond. Coulson (1941, 1947) has shown that the butadiene molecule may be represent as shown in the

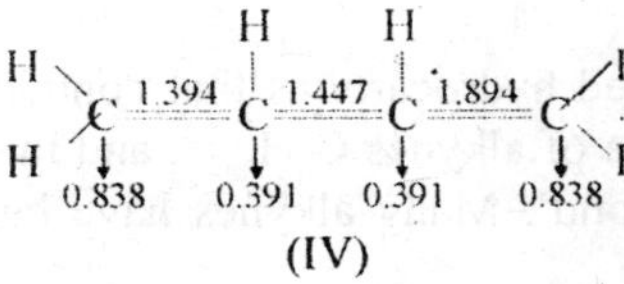

(IV) 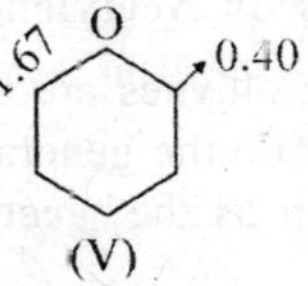 (V)

Calculation gives a bond order of 1.894 to the two outer 'double' bonds, and a bond order of 1.47 the central 'single' bond. Thus the total *bond number* of either of the end carbon atoms is 2 × 1.894 = 3.894, and the total bond number of either middle is 1.0 + 1.447 +1.894 = 4.34.

Furthermore, since calculation has shown that the maximum bond number for a carbon atomic 4.732, it follows that each carbon atom in butadiene has the free valency shown in (IV) (Coulson has suggested that free valency be represented by an arrow). On the other hand, if a structure containing fractional double bonds is written with single bonds labelled with the bond order and charges are placed on the atoms, then the resulting diagram is known as a *molecular diagram, e.g.*, the molecular diagram for benzene is (V).

More recently, it has been suggested that the relative stability of butadiene is not due toresonance stabilisation but is due to the variations in bond energies that arise from differences in the hybrid state of carbon (see p. 139). Table 2.1 gives some *experimental* bond lengths, and bond-energy values *calculated* by Bloor and Gartside (1959).

Table 2.1

Bond type	*Bond length (A)*	*Bond energy* kJ mol^{-1}	*(kcal mol^{-1})*
$C(sp^3)$—$C(sp^3)$	1.54	346.3	(82.76)
$C(sp^3)$—$C(sp^2)$	1.50	357.6	(85.48)
$C(sp^3)$—C(sp)	1.46	382.5	(91.42)
$C(sp^2)$—$C(sp^2)$	1.48	383.2	(91.58)
$C(sp^2)$—C(sp)	1.43	403.7	(96.48)
C(sp)—C(sp)	1.38	433.5	(103.6)
$C(sp^3)$—H	1.11	412.8	(98.67)
$C(sp^2)$—H	1.10	412.9	(98.69)
C(sp)—H	1.08	428.4	(102.38)
$C(sp^2)$ = C (sp^2)			

Alkynes or Acetylenes

The alkynes are unsaturated hydrocarbons that contains one triple bond (C°C) the general formula of alkynes C_nH_{2n-2} and the triple bond is known as the 'acetylenic bond'. Many alkynes have been found in nature.

Alkynes can be described by the condensed formula R – C ≡ C ≡ – R when R may be an alkyl group or H.

The simplest member of this homologous series (putting, n = 2) has the molecular formula C_2H_2.

$$H \overset{\sigma}{—} C \equiv C \underset{106\ pm}{\overset{\sigma}{—}} H \quad \text{or} \quad H—C\equiv C—H$$

one σ-bond + two π bonds

The triple bond present in alkynes – C ≡ C – is the functional group for alkynes. The properties of alkynes are largely determined by the triple bond (–C ≡ C –). In ethyne (or in alkynes in general), the triple-bonded carbon atoms exhibit *sp* hybridisation. Therefore, ethyne molecule is a linear molecule.

There are in all 3 σ-bonds and 2 π-bonds in a molecule of ethyne. The C ° C bond energy is 823 kJmol^{-1} – the bond energy of a C = C doubled bond is 678kJ mol^{-1}, and that of a C – C bond is 370kJ mol^{-1}.

Alkynes are named by replacing the suffix "ane" of the parent hydrocarbon by "ane". Some simple alkynes are,

$HC \equiv CH$ (ethyne) $CH_3 - C \equiv CH$ (propyne) $H_3C - C \equiv CH - CH_3$ (but-2-yne) $H_3CH_2 - C \equiv CH$ (but-1-yne)

Isomerism in Alkynes

Alkynes show three types of isomerism, viz.,

(a) Chain isomerism (b) Position isomerism (c) Functional isomerism

These are described below.

(i) *Chain isomerism :* It is due to the different arrangement of carbon atoms in the chain, *i.e.*, straight chain or a branched, *e.g.*

Branched chain: $CH_3—CH(CH_3)—C\equiv CH—CH_3$

4-methylpent-2-yne

Straight chain: $CH_3—CH_2—CH_2—C\equiv CH—CH_3$

hex-2-yne

(ii) *Position isomerism* : It is due to the difference in the location of the triple bond, *e.g.*,

$CH_3—CH_2—CH_2—C \equiv CH$ — pent-1-yne

$CH_3—CH_2—C \equiv C—CH_3$ — pent-2-yne

(iii) *Functional isomerism* : Alkynes are isomeric with alkadienes both being represented by the general formula C_nH2_{n-1}. For example

$CH_3—CH_2—C \equiv CH$ — but-1-yne

$CH_2 = CH—CH = CH_2$ — buta-1, 3-diene

ACIDIC CHARACTER OF HYDROGEN IN ETHYNE AND 1-ALKYNES

The hydrogen atoms in ethyne and 1-alkynes, linked to the carbon atom a having a triple bond on it are acidic in nature. For example, ethyne (acetylene) is a weak acid: weaker than water but stronger than ammonia. This may be explained as follows:

The π-electrons are more weakly bound than σ electrons. Thus, in compounds containing carbon-double bonds or triple bonds, the electron density around such carbon atoms is less than that on the carbon atoms linked only through s bonds. Thus, the electronegativity of differently hybridised carbon atoms follows the order,

$$sp > sp^2 > sp^3$$

i.e., the electronegativity increases with the s-character in the hybrid orbitals. This increase in the electronegativity of an alkyne carbon relative to the carbon atoms in alkenes and alkanes thus polarizes the C—H electron bond towards carbon and thereby facilitates the release of proton(s). Therefore, the acid strength of hydrogens follow the order,

$$\text{Alkynes} > \text{Alkenes} > \text{Alkanes}$$

The stabilities of the carbanions (anion left after the removal of proton) follow the order,

$$R—C \equiv C^- > R—CH = CH^- > R—CH_2—CH_2^-$$

Thus, the acid strength follows the order,

$$HC \equiv CH > C_2C = CH_2 > H_3C—CH_3$$

Compared to the organic acids, *e.g.*, ethanoic acid (CH_3COOH), ethyne is about 10^{20} time less acidic, while ethane is 10^{40} times less acidic.

PREPARATION OF ALKYNES

In contrast to alkanes and alkenes, not many methods are available fro the preparation of alkynes.

Alkynes may be prepared by any of the following methods.

FROM HALOGEN DERIVATIVES OF ALKANES

Alkynes can be prepared from *di-* and *tetra-* halogen derivatives of alkanes.

From Dihaloalkanes

Alkynes may be prepared any dehydrohalogenation (elimination) of dialoalkanes preferably bromo derivatives. Since there are two types of dihaloalkanes, hence the method of preparation from both are described below.

(a) *From alkylene halids (vic-dihalides) :* An alkylene dihalide has two halogen atoms: one each on the two adjacent carbon atoms. When boiled with alcoholic potash, the molecule loses 2 molecules of hydrogen halide and the corresponding alkyne is obtained.

$$\underset{\text{alkylene dihalide}}{R-\overset{X}{\underset{H}{C}}-\overset{H}{\underset{X}{C}}-R} \qquad R-\underset{H}{C}-\underset{X}{C}-R \xrightarrow[-KX,\ -H_2O]{KOH\ (alc)} \underset{\text{alkyne}}{R-C\equiv C-R}$$

For example, 1, 2-dibromoethane on boiling with alcoholic KOH gives ethyne.

$$\underset{\text{1,2-dibromoethane}}{\begin{matrix} CH_2Br \\ | \\ CH_2Br \end{matrix}} \xrightarrow[-KX,\ -H_2O]{KOH\ (alc)} \begin{matrix} CH_2 \\ \| \\ CHBr \end{matrix} \xrightarrow[-KX,\ -H_2O]{KOH\ (alc)} \underset{\substack{\text{ethyne} \\ \text{(acetlene)}}}{\begin{matrix} CH \\ ||| \\ CH \end{matrix}}$$

(b) *From alkylidene halides (gem-dihalides) :* The dihalo derivatives in which both the halogen atoms are attached to the same carbon atoms are known as *gem-dihalides.* When a gem-halides is boiled with alcoholic KOH corresponding alkyne is obtained. For example,

$$\begin{array}{c} CH_3 \\ | \\ CHBr_2 \end{array} + 2KOH\,(alc) \xrightarrow{\text{boiling}} \begin{array}{c} CH_3 \\ ||| \\ CHBr_2 \end{array} + 2KBr + 2H_2O$$

ethyne (acetylene)

From Tetrahaloalkanes

Alkynes can be obtained from tetrahaloalkanes by heating with zinc.

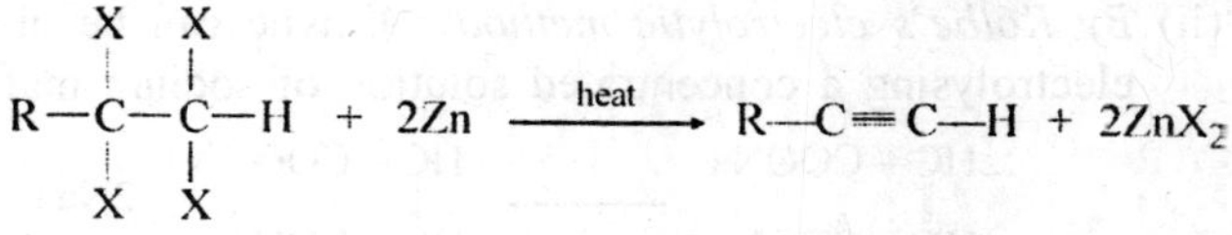

$$R-\overset{X}{\underset{X}{C}}-\overset{X}{\underset{X}{C}}-H + 2Zn \xrightarrow{\text{heat}} R-C\equiv C-H + 2ZnX_2$$

1,1,2,2-tetrahaloalkane alkyne

For example,

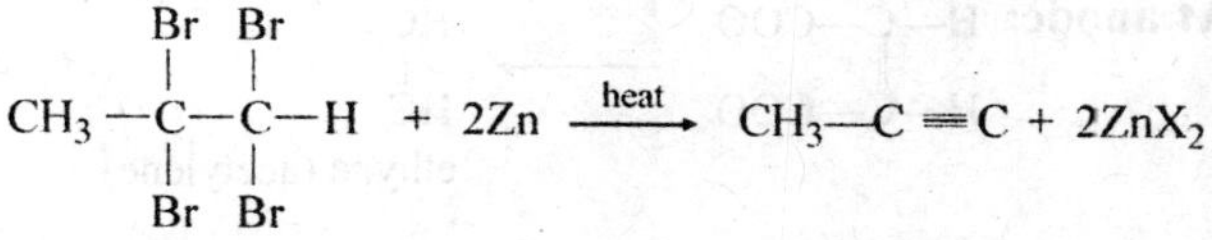

$$CH_3-\overset{Br}{\underset{Br}{C}}-\overset{Br}{\underset{Br}{C}}-H + 2Zn \xrightarrow{\text{heat}} CH_3-C\equiv C + 2ZnX_2$$

1,1,2,2-etrabromopane propyne

From Ethyne

Ethyne reacts with sodium amide (a mixture of Na in ammonia) to form sodium ethynide. The ethynide on treatment with an alkyl halide (haloalkane) leads to a higher alkyne.

$$\underset{\text{ethyne}}{HC\equiv CH} + NaNH_2 \xrightarrow{\text{ether}} \underset{\text{sodium ethynide}}{HC\equiv C^-Na^+} + NH_3$$

$$HC\equiv C^-Na^+ + \underset{\text{bromomethane}}{BrCH_3} \xrightarrow{\text{ether}} \underset{\text{bromomethane}}{HC\equiv C-CH_3}$$

$$\underset{\text{bromopropane}}{HC\equiv C^-\ NH^+} + \underset{\text{pent-1-yne}}{CH_3CH_2CH_2Br} \xrightarrow{\text{ether}} HC\equiv C-CH_2-CH_3 +$$

NaBr

Preparation of Ethyne (Acetylene)

Ethyne (acetylene) is the samples alkyne. It can be obtained by the following methods.

(i) *From calcium carbide* : Ethyne (acetylene) is obtained by reacting calcium carbide with water. Calcium carbide is obtained by heating limestone with coke

$$CaCO_3(s) + 3C\ (\text{graphite}) \xrightarrow{3000^\circ C} \underset{\text{calcium carbide}}{CaC_2(s)} + CO + CO_2$$

$$CaC_2 + 2H_2O \longrightarrow Ca(OH)_2 + \underset{\text{ethyne (acetylene)}}{C_2H_2}$$

This method is also used for industrial preparation of ethyne.

(ii) *By Kolbe's electrolytic method* : Ethyne can be obtained by electrolysing a concentrated solution of sodium maleate.

$$\underset{\text{sod. maleate}}{\begin{matrix} HC\text{—}COONa \\ \| \\ HC\text{—}COONa \end{matrix}} \longrightarrow \begin{matrix} HC\text{—}COO \\ \| \\ HC\text{—}COO \end{matrix} + 2Na^+$$

At anode:
$$\begin{matrix} H\text{—}C\text{—}COO^- \\ \| \\ H\text{—}C\text{—}COO^- \end{matrix} \longrightarrow \underset{\text{ethyne (acetylene)}}{\begin{matrix} HC \\ \| \\ HC \end{matrix}} + 2CO_3 + 2e^-$$

At cathode:
$$2H_2O \longrightarrow 2H^+ + 2OH^-$$
$$2H^+ + 2e^- \longrightarrow H_2(g)$$

Net reaction:
$$2H_2O + 2e^- \longrightarrow H_2(g) + 2OH^-$$

Thus, ethyne can be obtained at anode and hydrogen at cathode. This method is used only for preparing ethyne (acetylene).

PHYSICAL PROPERTIES OF ALKYNES

Some general physical properties of alkynes are described below:

(i) *Physical state:* First three members (ethyne, propyne and butyne) are colourless and odourless gases. Ethyne (acetylene) has garlic smell due to the presence of phosphine as impurity.

Next eight members are liquids, and higher members are solids under normal conditions of temperature and pressure.

(ii) *Solubility:* Alkynes are insoluble in water, but are fairly soluble in organic solvents such as, alcohol, ether acetone etc.

(iii) *Melting and boiling points:* The melting and boiling points of alkynes increase with molecular mass. The boiling points of some alkynes are listed below in Table 2.2.

Table 2.2 : Boiling Points of Some Alkynes

Alkyne		*Boiling point*	
Name	*Formula*	*°C*	*K*
Ehtyne (acetylene)	$CH \equiv CH$	–83	190
Propyne	$CH_3—C \equiv CH$	–27	246
But-1-yne	$CH_3—CH_2—C \equiv CH$	8	281
Pent-1-yne	$CH_3—C \equiv C—CH_3$	29	302
Pent-1-yne	$CH_3—CH_2—CH_2—C \equiv CH$	48	321
Pent-2-yne	$CH_3—C \equiv C — CH_2 — CH_3$	55	328

GENERAL CHEMICAL PROPERTIES OF ALKYNES

Alkynes contain a triple bond. A triple bond has one σ and two π bonds. Like alkenes, alkynes also show reactivity due to the presence of π bonds. As a result, alkynes given electrophilic addition reactions. However, alkynes are less reactive than alkenes towards electrophilic addition reactions due tothe compact $C \equiv C$ electron cloud.

Some characteristic reactions of alkynes are

Combustion

Alkynes burns in air or oxygen with smoky flame.

$$C_nH_{2n-2} + \frac{(3n-1)}{2} O_2 \longrightarrow nCO_2 + (n-1)\ H_2O$$

Electrophilic Addition Reactions

Carbon-carbon triple bond is a combination of one s and two p-bonds. *Due to the availability of electrons, alkynes given electrophilic addition reactions.* However, alkynes are less reactive than likeness towards electrophilic addition reactions. This is due to the compactness of π electron cloud in alkynes. Some typical electrophilic addition reactions given by alkynes are:

(i) *Addition of hydrogen:* An alkyne reacts with hydrogen in the presence of catalyst (Pt or Ni) at 250° C, first forming alkenes and finally alkane.

$$\underset{\text{alkyne}}{R—C\equiv CH} \xrightarrow[\text{Pt/Ni}]{+H_2} \underset{\text{alkene}}{R—C=CH_2} \xrightarrow[\text{Pt/Ni}]{+H_2} \underset{\text{alkane}}{R—CH_2=CH_3}$$

For example, ethyne gives ethane in two steps.

$$\underset{\text{ethyne}}{CH\equiv CH} + H_2 \xrightarrow{\text{Ni}} \underset{\text{ethene}}{CH_2=CH_2} \xrightarrow{+H_2/\text{Ni}} \underset{\text{ethane}}{CH_3—CH_3}$$

Propyne gives,

$$\underset{\text{propyne}}{CH_3—C\equiv CH} \xrightarrow[\text{Ni}]{+H_2} \underset{\text{propene}}{CH_3—CH=CH_2} \xrightarrow[\text{Ni}]{+H_2} \underset{\text{alkane}}{R—CH_2—CH_3}$$

If hydrogenation is done with a calculated amount of hydrogen in the presence of palladium on barium sulphate inquiline, *(Lindlar's catalyst),* it proceeds only upto alkene stage. Thus,

$$\underset{\text{ethyne}}{CH\equiv CH} + H_2 \xrightarrow[\text{quionoline}]{\text{pd/BaSO}_3} \underset{\text{ethene}}{CH_2=CH_2}$$

(ii) *Addition of halogens :* Alkynes react with halogens (Cl_2 or Br_2) in the dark, first forming dihaloalkens and finally tetrahaloalkanes. The reaction gets accelerated in the presence of light or halogen carriers.

$$\underset{\text{alkyne}}{R—C\equiv CH} \xrightarrow{+X_2} \underset{\text{dihaloalkane}}{R—CX\equiv CHX} \xrightarrow{+X_2} \underset{\text{tetrahaloalkane}}{R—CX_2\equiv CHX_2}$$

For example, ethyne (acetylene) with chlorine gives,

$$\underset{\text{ethyne}}{CH\equiv CH} + Cl_2 \longrightarrow \underset{\text{dichloroethene}}{CHCl=CHCl} \xrightarrow{+Cl_2} \underset{\text{tetrachloroethane}}{CHCl_2\equiv CHCl_2}$$

Dilute bromine water with ethyne gives dibromo, while liquid gives tetrabromo derivative.

$$\underset{\text{tetrabromoethane}}{CHBr—CHBr} \xleftarrow{2Br_2(l)} \boxed{\underset{\text{ethene (acetylene)}}{CH\equiv CH}} \xrightarrow[\text{dilute}]{+Br_2(aq)} \underset{\text{1,2-dibromethene}}{CHBr_2\equiv CHBr_2}$$

Propyne gives,

$$\underset{\text{propyne}}{CH_3—C\equiv CH} \xrightarrow{+Br_2} \underset{\text{1,2-dibromopropene}}{CH_3—C(Br)—CHBr} \xrightarrow{+Br_2} \underset{\text{1,1,2,2-tetrabromopropane}}{CH_3—CBr_2—CBr_2—CH}$$

The order of reactivity of halogens is $Cl_2 > Br_2 > I_2$.

(iii) *Addition of hydrogen halide:* Alkynes reacts with hydrogen halides (halogen acids) accordings to the *Markownikoff's rule — the negative part of the addendum gets attached to the carbon atom carrying the least number of hydrogen atoms.*

$$\underset{\text{alkyne}}{\begin{matrix} R\text{—}C \\ ||| \\ CH \end{matrix}} + \underset{\text{(halogen acids)}}{HX} \longrightarrow \underset{\text{2-haloalkene}}{\begin{matrix} R\text{—}C \\ || \\ CH_2 \end{matrix}} \xrightarrow{+\,HX} \underset{\text{2.2-cihaloalkane}}{\begin{matrix} R\text{—}CX_2 \\ | \\ CH_3 \end{matrix}}$$

For example, ethyne (acetylene) with HBr gives,

$$\underset{\substack{\text{ethyne}\\\text{(acetylene)}}}{\begin{matrix} CH \\ ||| \\ CH \end{matrix}} + HBr \longrightarrow \underset{\substack{\text{bromoethene}\\\text{(vinyl bromide)}}}{\begin{matrix} CHBr \\ || \\ CH_2 \end{matrix}} \xrightarrow{+\,HBr} \underset{\substack{\text{1,1-dibromoethane}\\\text{(ethylidene dibromide)}}}{\begin{matrix} CHBr_2 \\ | \\ CH_3 \end{matrix}}$$

With dil. HCl at 65° C and in the presence of Hg^{2+} (mercuric ion) ethyne gives vinyl chloride.

$$\underset{\text{ethyne}}{\begin{matrix} CH \\ ||| \\ CH \end{matrix}} + HCl \xrightarrow{65°C,\ Hg^{2+}} \underset{\text{vinyl chloride}}{\begin{matrix} CH_2 \\ || \\ CHCl \end{matrix}}$$

Propyne gives,

$$\underset{\text{propyne}}{\begin{matrix} CH_3\text{—}CH \\ ||| \\ CH \end{matrix}} + Br \longrightarrow \underset{\text{2-bromopropene}}{\begin{matrix} CH_3\text{—}CBr \\ || \\ CH_2 \end{matrix}} \xrightarrow{+HBr} \underset{\text{2,2-dibromoprpane}}{\begin{matrix} CH_3\text{—}CBr \\ | \\ CH_3 \end{matrix}}$$

The rate of addition of halogen acids follows the order,

$$HI > HBr > CHI$$

Mechanism

$$R\text{—}C\equiv CH \longrightarrow R\text{—}C^{+}=\ddot{C}H^{-} \xrightarrow{H^{\delta+}Br^{\delta-}} R\text{—}\underset{Br}{\underset{|}{C}}=\underset{H}{\underset{|}{CH}}$$

(iv) *Addition of hypochlorous acid:* Alkynes react with hypochlorous acid according to the Markownikoff's rule

$$R\text{—}\underset{Br}{\underset{|}{C}}\equiv CH_2 \longrightarrow R\text{—}\underset{Br}{\underset{|}{C^{+}}}=\ddot{C}H^{-} \xrightarrow{H^{\delta+}Br^{\delta-}} R\text{—}\overset{Br}{\overset{|}{\underset{Br}{\underset{|}{C}}}}=CH_3$$

For example, ethyne with HOCl gives,

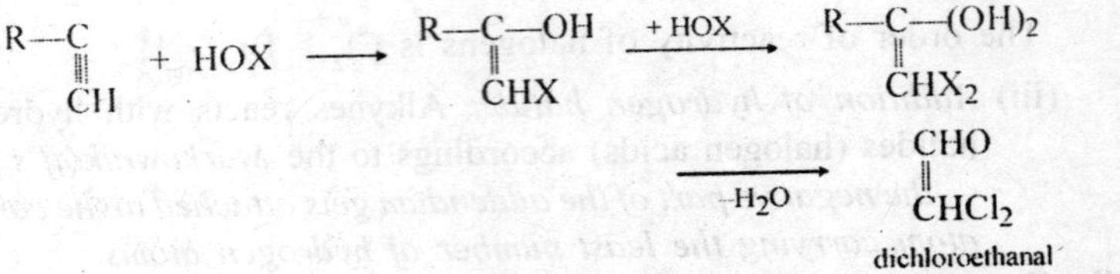

$$R{-}C{\equiv}CH + HOX \longrightarrow R{-}C(OH){=}CHX \xrightarrow{+HOX} R{-}C(OH)_2{-}CHX_2 \xrightarrow{-H_2O} CHO{-}CHCl_2$$

dichloroethanal

In the presence of peroxides, the addition of HBr takes place according to the anti-Markownikoff's rule.

(v) *Addition of sulphuric acid.* Alkynes add up two molecules of sulphuric acid. For example, ethyne gives

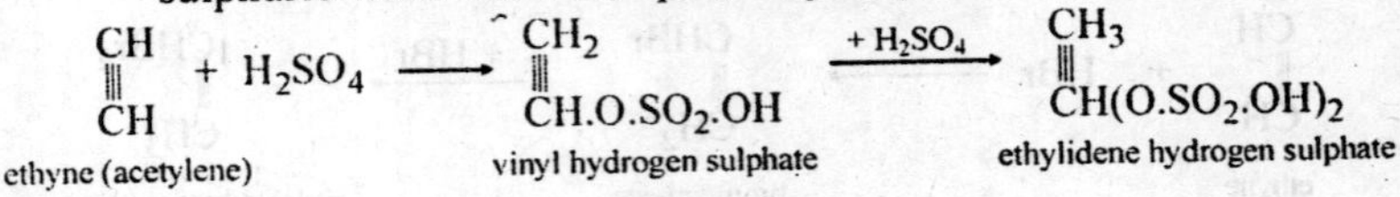

$$CH{\equiv}CH + H_2SO_4 \longrightarrow CH_2{=}CH.O.SO_2.OH \xrightarrow{+H_2SO_4} CH_3{-}CH(O.SO_2.OH)_2$$

ethyne (acetylene) — vinyl hydrogen sulphate — ethylidene hydrogen sulphate

Nucleophilic Addition Reactions

Alkynes also give the following nucleophilic addition reactions.

(i) *Addition of water :* Alkynes in the presence of sulphuric acid (42%) and 1% mercuric sulphate at 60°C, add one water molecule to give aldehydes or ketones. For example,

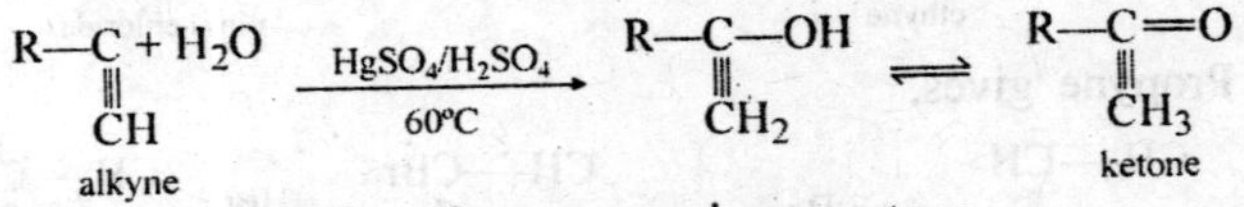

$$R{-}C{\equiv}CH + H_2O \xrightarrow[60^{\circ}C]{HgSO_4/H_2SO_4} R{-}C(OH){=}CH_2 \rightleftharpoons R{-}C({=}O){-}CH_3$$

alkyne — ketone

Ethyne gives ethanol and propyne gives acetone.

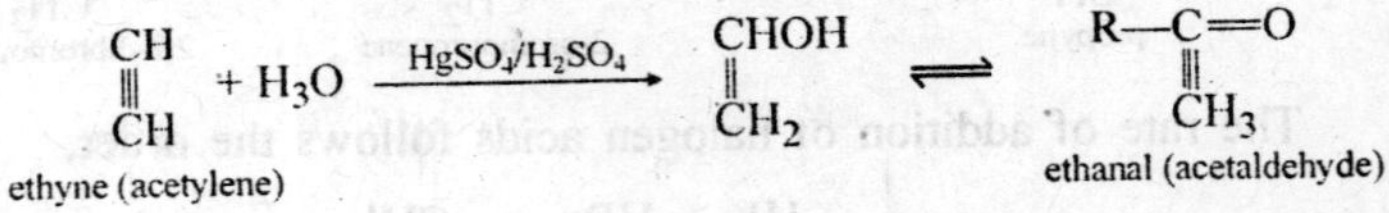

$$CH{\equiv}CH + H_3O \xrightarrow{HgSO_4/H_2SO_4} CHOH{=}CH_2 \rightleftharpoons R{-}C({=}O){-}CH_3$$

ethyne (acetylene) — ethanal (acetaldehyde)

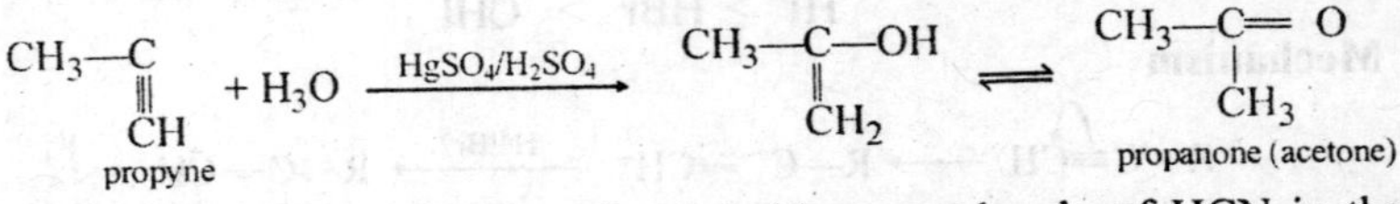

$$CH_3{-}C{\equiv}CH + H_3O \xrightarrow{HgSO_4/H_2SO_4} CH_3{-}C(OH){=}CH_2 \rightleftharpoons CH_3{-}C({=}O){-}CH_3$$

propyne — propanone (acetone)

(ii) *Addition of HCN :* Alkynes add one molecule of HCN in the presence of $Ba(CN)_2$. For example,

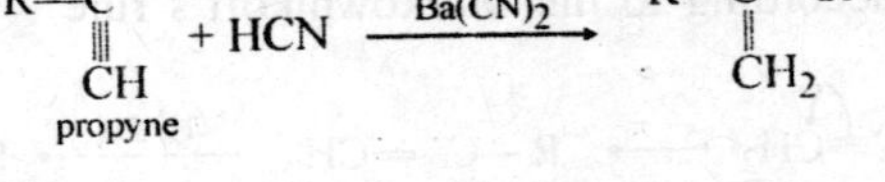

$$R{-}C{\equiv}CH + HCN \xrightarrow{Ba(CN)_2} R{-}C(CN){=}CH_2$$

propyne

Ethyne gives,

$$CH{\equiv}CH + HCN \xrightarrow{Ba(CN)_2} CH_2{=}CHCN$$

ethyne — vinyl cyanide

(iii) *Addition of alcohols :* Ethyne adds a molecule of alcohol in the presence of alkali to give vinyl ether.

$$\underset{\text{ethyne}}{CH\equiv CH} + \underset{\text{methanol}}{H_3COH} \xrightarrow[393\text{-}453\ K]{KOH} \underset{\substack{\text{methyl vinyl ether}\\ \text{(methoxyethene)}}}{H_3CO{-}CH{=}CH_2}$$

(iv) *Addition of acids :* Alkynes add acids in the presence of Lewis acid catalyst or Hg^{2+} ions to form vinyl esters.

$$H{-}C\equiv CH + C_3CCOOH \xrightarrow{Hg^{3+},\ 353\ K} \underset{\text{vinylethonate}}{CH_3{-}\overset{\overset{O}{\|}}{C}{-}O{-}CH = CH_2}$$

$$R{-}C\equiv CH + C_3CCOOH \xrightarrow[\text{(Lewis acid)}]{BF_3} \underset{\alpha\text{-alkylvinyl ethonate}}{CH_3{-}\overset{\overset{O}{\|}}{C}{-}O{-}\overset{\overset{R}{|}}{C} = CH_2}$$

(v) *Addition of ozone:* Ozone adds up across the triple bond to give ozonides. Ozonides on hydrolysis give diketones and carboxylic acids.

$$\underset{\text{alkyne}}{R{-}C\equiv CR'} + O_3 \longrightarrow \underset{\text{ozonide}}{R{-}C(O)(O{-}O)C{-}R'} \xrightarrow{H_2O} R{-}\underset{\underset{O}{\|}}{C}{-}\underset{\underset{O}{\|}}{C}{-}R' + H_2O_2$$

$$\longrightarrow RCOOH + R'COOH$$

Ethyne gives glyoxal and then formic acid.

$$\underset{\text{alkyne}}{CH\equiv CH} + O_3 \longrightarrow \underset{\text{ozonide}}{HC(O)(O{-}O)CH} \xrightarrow{H_2O} H\underset{\underset{O}{\|}}{C}{-}\underset{\underset{O}{\|}}{C}H + H_2O_2 \longrightarrow \underset{\text{formic acid}}{2HCOOH}$$

Substitution Reactions

Alkynes due to their acidic nature, form metallic salts, *e.g.*, socium silver and copper (ous) salts. The metallic salts of alkynes are callec *alkynides*. For example,

$$\underset{\text{alkyne}}{R{-}C\equiv CH} + Na(\text{liq. } NH_3) \longrightarrow R{-}C\equiv CNa + NH_4^+$$

$$R{-}C\equiv CH + Ag(NH_3)_2^+ \longrightarrow \underset{\text{white ppt.}}{R{-}C\equiv CAg} + NH_4^+$$

$$R{-}C\equiv CH + Cu(NH_3)_2^+ \longrightarrow \underset{\text{red ppt.}}{R{-}C\equiv CCu} + NH_4^+ + NH_3$$

Ethyne (acetylene) has two acidic hydrogen atoms, hence it finally gives dimetal salts.

$$\underset{\substack{\text{disilver}\\ \text{ethynide}\\ \text{(white ppt.)}}}{CAg \equiv CAg} \xleftarrow{Ag^+} \underset{\substack{\text{monosilver}\\ \text{ethynide}}}{CH \equiv CAg} \xleftarrow{Ag^+} \boxed{CH \equiv CH} \xrightarrow{Na(in\ NH_3)} \underset{\substack{\text{monosodium}\\ \text{ethynide}}}{CH \equiv CNa} \xrightarrow{Na} \underset{\substack{\text{disodium}\\ \text{ethynide}}}{CNa \equiv CNa}$$

$$CH \equiv CH \xrightarrow{Cu^+} CH \equiv CCu \xrightarrow{Cu^+} \underset{\text{dicopper ethynide (red ppt)}}{CCu \equiv CCu}$$

Polymerisation

Alkynes on heating in the presence of catalyst undergo polymerisation. The nature of products depends upon the conditions. For example,

(i) When ethyne (acetylene) is passed through a hot copper tube, it polymerises to benzene.

$$3\ CH \equiv CH \xrightarrow{\text{hot cu tube}} \underset{\text{benzene}}{[HC \equiv CH_3,\ CH \equiv CH,\ HC \equiv CH]} \longrightarrow C_6H_6 \text{ or } \text{benzene}$$

(ii) Ethyne undergoes linear polymerisation when passed through a solution of cuprous chloride in ammonium chloride.

$$2\ CH \equiv CH \xrightarrow{Cu_2Cl_2/NH_4Cl} \underset{\text{vinylacetylene}}{CH_2 = CH—C \equiv CH} \xrightarrow{+CH \equiv CH} \underset{\text{divinylacetylene}}{CH_2 = CH—C \equiv C—CH = CH_2}$$

Polymerisation of acetylene (ethyne) under certain suitable conditions, gives polyacetylene (polyethyne).

$$\underset{\substack{\text{acetylene}\\ \text{(ethyne)}}}{2n\ HC \equiv CH} \longrightarrow \underset{\substack{\text{polyacetylene}\\ \text{(polyethyne)}}}{-(CH = CH—CH = CH)_n}$$

Polyacetylene is a high molecular mass, conjugated polymer. It is an electrically conducing polymer. Thin films of polyacetylene are used for making electrodes for batteries. Polymers such as polyacetylene hold a very bright future.

(iii) Vinyl acetylene with hydrogen chloride gives chloroprene (2-chlorobuta-1, 3-diene), which readily polymerises to give neoprene (a synthetic rubber).

$$\underset{\text{vinylacetylene}}{CH_2=CH—C\equiv CH} + HCl \longrightarrow \underset{\substack{\text{chloroprene}\\ \text{(2-chlorobuta-1,3-diene)}}}{CH_2=CH—\underset{|}{\overset{}{C}}(Cl)=CH_2}$$

Oxidation

Oxidation of alkynes give mono or dicarboxylic acids. For example,

(i) Alkaline $KMnO_4$ oxidises ethyne to oxalic acid.

$$\underset{\text{ethyne}}{CH\equiv CH} + O_2 \longrightarrow \underset{\substack{\text{oxalic acid}\\ \text{(ethanedioic acid)}}}{COOH—COOH}$$

(ii) With chromic acid, ethyne gives acetic acid.

$$\underset{\substack{\text{ethyne}\\ \text{(acetylene)}}}{CH\equiv CH} + [O] + H_2O \xrightarrow{H_2CrO_4} \underset{\substack{\text{ethanoic acid}\\ \text{(acetic acid)}}}{CH_3—COOH}$$

(iii) Homologues of ethyne of oxidation with alkaline $KMnO_4$ give mixture of acids. During oxidation, the rupture takes place at the triple bond.

$$\underset{\text{alkyne}}{R—C\equiv C—R'} + (O) + H_2O \xrightarrow{KMnO_4\,(alk)} RCOOH + R'COOH$$

Uses of Alkynes

Alkynes are generally used as the starting materials for the manufacture of a large number of organic compounds of industrial importance such as, chloroprene, vinyl chloride etc.

TESTS FOR DETECTING UNSATURATION IN HYDROCARBONS

Hydrocarbons are termed unsaturated, if they contain one or more double and/or triple bonds in their molecules. The presence of a double and/or triple bond (s) can be confirmed by the following test.

(i) *Bromine test* : Add 2 ml of bromine solution in carbon tetrachloride to the jar containing the hydrocarbon and shake. The decolorization of bromine colour confirms the presence of a double or/and triple bond in the molecule of the hydrocarbon.

This decolorisation takes place due to the addition of bromine across the double or triple bond.

$$\underset{\text{alkene}}{>C=C<} \quad + \quad \underset{\text{reddish orange}}{Br_2(\text{in } CCl_4)} \longrightarrow \underset{\text{colourless}}{>\overset{\overset{Br}{|}}{C}=\overset{\overset{Br}{|}}{C}<}$$

$$\underset{\text{alkene}}{>C\equiv C-} \quad + \quad \underset{\text{reddish orange}}{Br_2(\text{in } CCl_4)} \longrightarrow \underset{\text{colourless}}{>\overset{\overset{Br}{|}}{C}=\underset{\underset{Br}{|}}{C}—} \longrightarrow \underset{\text{colourless}}{—\overset{\overset{Br}{|}}{\underset{\underset{Br}{|}}{C}}=\overset{\overset{Br}{|}}{\underset{\underset{Br}{|}}{C}}—}$$

(ii) *Bayer's test ($KMnO_4$ test)* : Alkaline $KMnO_4$ solution is called *Bayer's reagent*. Alkenes and alkynes (unsaturated hydrocarbons) decolorise the pink colour of $KMnO_4$ (alkaline). So, if the colour of alkaline is decolorised, the test sample should contain any unsaturated hydrocarbon.

$$\underset{\text{alkene}}{>C=C<} + \underset{\substack{\text{from alkaline}\\ KMnOr_4\text{, pink}}}{H_2O+[O]} \longrightarrow \underset{\text{alkanedio}}{>\overset{\overset{Br}{|}}{C}—\overset{Br}{C}<}$$

$$\underset{\text{alkene}}{R—C\equiv C—R'} + \underset{\substack{\text{from alkaline}\\ KMnOr_4\text{, pink}}}{H_2O+3[O]} \longrightarrow \underset{\substack{\text{mixture of carboxylic a}\\ \text{cids, colourless}}}{RCOOH + R'COOH}$$

TESTS FOR THE PRESENCE OF A TRIPLE BOND

The presence of a triple bond in any hydrocarbon makes it unsaturated, and the compound gives the test described above. However, alkynes (compounds containing triple bonds) are characterised by certain specific tests described below.

(i) *With ammoniacal silver nitrate:* Alkynes give white precipitate of silver acetylide with ammoniacal silver nitrate.

(ii) *With ammonical cuprous chloride:* Alkynes give red precipitate of cuprous acetylide with ammoniacal cuprous chloride.

Distinction Between Ethane, Ethene and Ethyne

Ethane, ethene and ethyne are typical alkane, alkene and alkyne respectively. Some main differences between these three are given belcw.

Table 2.3

Property	*Ethane*	*Ethene*	*Ethyne*
Molecular parameters			
Common name	Ethane	Ethylene	Acetylene
Molecular formula	$C_2H_6C_2H_4$	C_2H_2	
Structural formula	H_3C-CH_3 (H—C(H)(H)—C(H)(H)—H)	$H_2C=CH_2$	H—C≡C—H
Molar mass	30 g mol^{-1}	28 g mol^{-1}	26 g mol^{-1}
Physical properties			
Physical state	Gas Gas	Gas	
Functional group	—	Double bond	Triple bond
Boiling point	–89ºC	–103.8º C	–84ºC
Meting point	–172ºC	–169.4ºC	—
Chemical properties			
(a) Burning in air sooty	Burns with non luminous flame	Burns with luminous flame	Bsurns with flame
(b) Nature of reactions addition	Substitution	Addition (mainly)	Mainly
(c) With alkaline $KMnO_4$	No action	Decolourises pink colour of $KMnO_4$	Decolourises pink colour of $KMnO_4$
(d) With bromine water	No action	Decolourises bromine water	Decolcurises bromine water
(e) Ammonical $AgNO_3$	No action	No action	White precipitate of silver acetylide
(f) Ammoniacal cuprous chloride	No action	No action	Red precipitate of copper acetylide

This order of reactivity is observed for catalytic hydrogenation (Reaction 1, below). The mechanism of this reaction, however, does not involve electrohilies. A possible contributing factor for decreased reactivity of a triple bond to electrophiles may be the fact that the bridged carbonium ion from acetylene (assuming its formation) is more strained than the bridged carboniumion from ethylene (the broken lines represent the bonds actually present in alkenes and alkanes):

$$—C\equiv C— \longrightarrow \overset{Br^+}{C \equiv C} \qquad C = C \longrightarrow \overset{Br^+}{C — C}$$

Justification for the assumption of these intermediate bridged ions being involved in addition to acetylene by the polar mechanism lies in the fact that the additions are stereoselectively *trans* (cf. ethylene).

Reaction of Acetylene

1. Acetylene adds on gaseous chlorine or bromine in the dark to form acetylen di-and tetrahalides; the addition is catalysed by light and metallic halides (*cf.* alkenes):

$$C_2H_2 \xrightarrow{Cl_2} C_2H_2Cl_2 \xrightarrow{Cl_2} C_2H_2Cl_4$$

Direct combination of acetylene with chlorine may be accompanied by explosions, but this is prevented by the presence of a metal chloride catalyst.

Acetylene reacts with dilute bromine water to produce acetylene dibromide:

$$CH \equiv CH + Br_2 \xrightarrow{H_2O} CHBr = CHBr$$

With liquid bromine and in the absence of a solvent, acetylene forms acetylene tetrabromide:

$$C_2H_2 + 2Br_2 \longrightarrow C_2H_2Br_4$$

Acetylene adds on iodine with difficulty, but if the reaction is carried out in ethanolic solution, acetylene di-iodide is obtained:

$$CH \equiv CH + I_2 \xrightarrow{ethanol} CHI = CHI$$

The addition of halogens to acetylene is stereoselective; the predominant product is the *trans* isomer (*cf.* the alkenes).

$$H—C\equiv C—H + X_2 \longrightarrow \begin{matrix} H & & X \\ & C=C & \\ X & & H \end{matrix}$$

Acetylene also undergoes substitution with halogen provided the right conditions are used, *e.g.*, dichloroacetylene is formed when acetylene is passed into sodium hypochlorite solution at 0° C in the absence of air and light:

$$H{-}C{\equiv}C{-}H \quad O{-}Cl \rightleftharpoons H{-}C{\equiv}C \quad Cl{-}O{-}H$$

$$\longrightarrow OH^- + H{-}C{\equiv}C{-}Cl \xrightarrow{OCl^-} Cl{-}C{\equiv}C{-}Cl$$

Since ethylene does not behave in a similar manner, it is reasonable to assume that it is the greater acidity of acetylene that enables this reaction to occur.

Similarly, if acetylenes passed into a solution of iodine in liquid ammonia, di-iodoacetylene is formed (Vaughan and Nieuland, 1932):

$$C_2H_2 + 2I_2 + 2NH_3 \longrightarrow C_2I_2 + 2NH_4\,I$$

2. Acetlene adds on hydrogen in the presence of a catalyst, the reaction proceeding in two stages:

$$C_2H_2 \xrightarrow[\text{cat.}]{H_2} C_2H_4 \xrightarrow[\text{cat.}]{H_2} C_2H_6$$

Most catalysts effect hydrogenation of a triple bond, and it is possible, by using suitable conditions, is isolate the intermediate alkene. One method uses Lindlar's catalyst (1952), which consists of a Pd—$CaCO_3$ catalyse partially poisoned with lead acetate. A better catalyst is Pd—$BaSO_4$ partially posioned with in line (Cram *et al.,* 1956)

Dialkylacetyles may be catalytically reduced to mixture of *eis-* and *trans-* alkenes, the former predominating. On the other hand, reduction with sodium in liquid ammonia produces the *trans*-alkene, which is also produced by reduction with lithium aluminiumhydride (Slaugh, 1966) cis-Reduction may be carried out with diborane, or with di-isobutylauminium hydride (Wike *et* al, 1960).

$$R{-}C{\equiv}C{-}R \longrightarrow \underset{cis}{(R)(H)C{=}C(R)(H)} \quad \underset{trans}{(R)(H)C{=}C(H)(R)}$$

3. Acetylene can add on the halogen acids, their order of reactivity being HI > HBr > HCl > HF; HF adds on only under pressure

(*cf.* ethylene). The addition of the halogen acids can take place in the dark but is catalysed by light or metallic halides. The addition is in accordance with Marownikoff's rule, *e.g.*, acetylene combines with hydrogen bromide to form first vinyl bromide, and then ethylidene dibromide:

$$CH \equiv CH + Br \longrightarrow CH_2 = CHBr \xrightarrow{HBr} CH_3CHBr_2$$

Peroxide have the same effect on the addition of hydrogen bromide to acetylene as they have on alkenes.

The mechanism of the addition of halogens and halogen acids is the same as that for the alkenes, *e.g.*, the addition of hydrogen bromide.

$$CH \equiv CH \quad H—Br \longrightarrow \bar{Br} \quad \overset{+}{C}H = CH_2 \longrightarrow CHBr = CH_2$$

Addition of another molecule of hydrogen bromide could give either the intermediate carbonium ion, $CH_3\overset{+}{C}HBr$ (a secondary ion), or $\overset{+}{C}H_2CH_2Br$ (a primary ion). Since the former is more stable than the latter, the reaction proceeds via the former to form ethylidene dibromide. Thus:

$$Br—H \quad CH_2 = CHBr \longrightarrow CH_3\overset{+}{C}HBr \quad \bar{Br} \longrightarrow CH_3CHBr_2$$

Because of the —1 effect of the bromine atom, the availability of the π-electrons in the first step in decreased and the addition is much slower than that for ethylene.

$$\overset{+}{C}H_2—\overset{-}{C}HBr \xrightarrow{-1} CH2 = CH—Br \xleftrightarrow{+R} \overset{-}{C}H_2—CH = \overset{+}{B}r$$

It might also be noted here that the contribution of the charged structure to the resonance hybrid, vinyl bromide, is not sufficient to give the bromine atom a positive charge. The direction hybrid, vinyl bromide, is not sufficient to give the bromine atom a positive charge. The direction of the dipole is still $\overrightarrow{C — BR}$,but its value is less than that in ethyl bromide. This observation, however, can also be explained in terms of the sp^2 hybridsation of the carbon atom (=CHBr): its electronegativity is greater than that of an sp^3 carbon atom.

4. When passed into dilute sulphuric acid at 60° C in the presence of mercuric sulphate as catalyst, acetylene adds on one moleeule of water to form acetaldehyde. The mechanism of the hydration probably takes place via the formation of vinyl alcohol as an intermediate.

$$CH \equiv CH + H_2O \xrightarrow[Hg^{2+}]{H_2SO_4} [CH_2 = CHOH] \longrightarrow CH_3CHO$$

The function of the catalyst is not certain. A possible explanation is that the mercuric ion, because of its large size, can fairly readily form a bridged ion (or a π-complex). When this is formed, reaction then proceeds as follows:

$$HC \equiv CH + Hg^{2+} \longrightarrow \underset{Hg^{2+}}{HC \equiv CH} \xrightarrow{H_2O} \underset{Hg^{+}}{HC = C\overset{+}{H}OH_2}$$

$$\xrightarrow{-H^+} \underset{Hg^{+}}{HC = CHOH} \xrightarrow{H_2O^-} Hg2+ + [CH_2{=}CHOH] \longrightarrow CH_3CHO$$

The homologues of acetylene form ketones when hydrated, *e.g.*, propyne gives acetone:

$$CH_3C \equiv CH + H_2O \xrightarrow[Hg^{2+}]{H_2SO_4} [CH_3C(OH) = CH_2] \longrightarrow CH_3COCH_3$$

5. When acetylene is passed into dilute hydrochloric acid at 65°C in the presence of mercuric ions as catalyst, vinyl chloride is formed:

$$CH \equiv CH + HCl \xrightarrow{Hg2^+} CH_2 = CHCl$$

Acetylene adds on hydrogen cyanide in the presence of cuprous chloride in hydrochloric acid as catalyst to form vinyl cyanide:

$$CH \equiv CH + HGN \longrightarrow CH_2 = CHCN$$

Vinyl cyanide is manufactured by passing a mixture of propene, ammonia, steam and air over a catalyst, *e.g.*, a mixture of oxides of molybdenum, cobalt and aluminium.

$$2CH_2 = CHCH_3 + 2NH_3 + 3O_2 \longrightarrow 3CH_2 = CHCN + 6H_2O$$

Vinyl cyanide is used in the manufacture of *Buna N* synthetic rubber, which is a copolymer of vinyl cyanide and butadiene.

When acetylene is passed into warm acetic acid in the presence of mercuric ions as catalyst vinyl acetate and ethylidene diacetate are formed:

$$CH \equiv CH + CH_3CO_2H \xrightarrow{Hg^{2+}} CH_2 = CHOOCCH_3$$

$$CH_2 = CHOOCCH_3 + CH_3CO_2H \xrightarrow{Hg^{2+}} CH_3CH(OOCH_3)_2$$

Vinyl acetate is manufactured by passing a mixture of acetylene and acetic and vapour over zinc acetate on charcoal at about 170°C, or by the reaction between ethylene, palladium chloride and sodium acetate in aqueous solution in the presence of cupric chloride.

$$CH_2 = CH_2 + PdCl_2 + 2CH_3CO_2\,Na$$

$$\xrightarrow{Ca^{2+}} CH_2 = CHOOCCH_3 + 2NaCl + Pd + CH_3CO_2H$$

Vinyl acetate (liquid) is used in the plastics industry. Ethylidene diacetate (liquid), when heated rapidly to 300-400° C, gives acetic anhydride and acetaldehyde.

Acetylene reacts with nitric acid in presences of mercuric ions to form nitroform, CH $(NO_3)_3$, and combines with aresenic trichloride to form Lewisite. When acetylene is passed into methanol at 160-200° C in the presence of a small amount (1-2 percent) of potassium methoxide and under pressure just high enough to prevent boiling, methyl vinyl ether is formed. The mechanism is believed to involve nucelophilic attach in the first site.

$$CH \equiv CH + CH_3O^- \longrightarrow \bar{C}H = CHOCH_3$$

$$\xrightarrow{CH_3OH} CH_2 = CHOCH_3 + CH_3\,O^-$$

Methyl vinyl either is used for making the polyvinyl ether plastics.

This process whereby acetylene adds onto compounds containing an active hydrogen atom to form vinyl compounds is known as vinylation.

Acetylene and formaldehyde interact in the presence of sodium alkoxide as catalyst to form butynediol, together with smaller amounts of propargyl alcohol:

$$HC \equiv CH + CH_3\,\bar{O} \rightleftharpoons HC \equiv C^- + CH_3OH$$

$$O{=}CH_2 \quad ^-C{\equiv}CH \longrightarrow {}^-O{-}CH_2{-}C \equiv CH \xrightarrow{CH_3OH}$$

$$OH{-}CH_2{-}C{\equiv}CH \xrightarrow[CH_2O]{CH_3OH} HOCH_2C{\equiv}CCH_2OH$$

This reaction in which acetylene (or any compound containing the ≡CH group *i.e.*, a methane hydrogen atom) adds on to certain unsaturated links (such as in the carbonyl group), or eliminates a molecule of water by reaction with certain hydroxy-compounds, is known as ethyinylation. Thus the above reactions with form aldehyde are examples of ethinylation; another example, is the following:

$$R_2NCH_2OH + CH \equiv CH \longrightarrow R_2NCH_2C \equiv CH + H_2O$$

Alkyl bromo and chloro-methyl ethers add to acetylenes in the presence of the corresponding aluminium halide to form alkenes (Bindacz et al., 1960); *e.g.*,

$$EtOCH_2Cl + CH \equiv CH \xrightarrow{AlCl_3} EtOCH_2CH = CHCl$$

Under the same conditions acyl chlorides behave similarly;

$$RCOCl + CH \equiv CH \xrightarrow{AlCl_3} RCOCH = CHCl$$

6. When acetlene is passed into hypochlorous acid solution, dichloroacetaldehyde is formed:

$$CH \equiv CH + HOCl \longrightarrow [CHCl = CHOH] \xrightarrow{HOCl} [Cl_2CHCH(OH)_2]$$

$$\longrightarrow Cl_2CHCHO + H_2O$$

Dichloroacetic acid, Cl_2CHCO_2H, is also formed by the oxidation of dichloroacetaldehyde by the hypochlorous acid.

7. Acetylene and its homologues form ozonides with ozone, and these compounds are decomposed by water to form diketones, which are then oxidised to acids by the hydrogen peroxide formed in the reaction:

$$HC \equiv CR^2 + O_3 \longrightarrow R^1C(-O-)CR^2 \text{ (ozonide with O—O bridge)} \xrightarrow{H_2O} R^1C(=O)-C(=O)R^2 + H_2O_2$$

$$\longrightarrow R^1CO_2H + R^2CO_2H + H_2O_2$$

Acetylene is exceptional in that it gives glyoxal as well as force acid (Hurd and Christ, 1936).

$$CH \equiv CH \xrightarrow[(ii)\,H_2O]{(i)\,O_3} OCHCHO + HCO_2H$$

Since oxidation (ozonide formation being one example) is very much slower with alkynes than with alkenes, by choosing a suitable oxidising agent, a triple bond may be left intact in enynes, *e.g.*,

$$R^1C \equiv C(CH_2)_n\, CH = CHR^2 \xrightarrow[AcOH]{CrO_3} R^1C \equiv C(CH_2)CO_2H + R^2COH$$

Hydrogen peroxide-formic acid mixture also preferentially hydroxylates the double bond in enynes. On the other hand, permanganate attacks a triple bond, *e.g.*,

$$CH_3(CH_2)_7C \equiv C(CH_2)_7CO_2H \xrightarrow{KMnO_4} CH_3(CH_2)_7COCO(CH_2)_7CO_2H$$

8. When passed through a heated tube acetylen polymerises, to a small extent, to benzene.

$$3C_2H_2 \longrightarrow \text{(benzene)}$$

Homologues of acetylene behave in a similar manner, *e.g.*, propyne polymerises to 1, 3, 5-trimethyl-benzene, and but-2-yne to hexamelthyl benzene:

$$3C_2H_2 \longrightarrow \text{(1,3,5-trimethylbenzene: } CH_3, CH_3, CH_3\text{)} \qquad 3C_3HC \equiv \longrightarrow \text{(hexamethylbenzene: } CH_3 \times 6\text{)}$$

Under suitable conditions acetylene polymerises to cyclooctatetraene (q.v).

$$4C_2H_2 \longrightarrow \text{(cyclooctatetraene)}$$

In addition to the above type of *cyclic* polymerisation, acetylene undergoes *linear* polymerisation when passed into a solution of cuprous chloride in ammonium chloride to give vinyalcetylene and divinylacetylene:

$$CH \equiv CH + CH \equiv CH \longrightarrow CH_2 = CH—C$$

$$\xrightarrow{C_2H_2} CH_2 = CH—C \equiv CH = CH_2$$

Compounds containing both a double and triple bond are named systematically as *alkenynes.* The double bond is always expressed first in the name, and numbers as low as possible are given to the double and triple bonds, even though this may given 'yne' the lower number. Thus vinylacetylene is but-1-en-3-yne and divinylacetylene is hexa-1, 5-dien-3-yne. The following compound, $CH_3CH = CHC = CH$, however, is pent-3-en-lyne.

$$CH_2 \equiv CH—C \equiv CH + HCl \longrightarrow [CH_2Cl—C = CH_2]$$

$$\longrightarrow CH_2 = CCl—CH = CH_2$$

Chloroprene readily polymerises to a rubber-like substance known as *neoprene.*

9. Acetylene forms metallic derivatives by replacement of one or both hydrogen atoms, *e.g.*, if acetylene is passed over heated, sodium both the monosodium and disodium acetylides are formed:

$$CH \equiv CH \xrightarrow{Na} CH \equiv CNa \xrightarrow{Na} NaC \equiv CNa$$

By using a large excess of acetylene the main product is monosodium acetylide, which is also obtained by passing acetylene into a solution of sodium in liquid ammonia until the blue colour disappears. BNy treatment of a fine dispersion of sodium in xylene at 100-105° C with acetylene, sodium acetylide can be obtained in 98 percent yield (Rutledge, 1957). The monosodium derivative possesses the interesting property of being able to absorb dry carbon dioxide to form the sodium salt of perioplic acid (q.v.):

$$CH \equiv CNa + CO_2 \longrightarrow CH \equiv CCO_2Na$$

The alkali metal acetylides react with carbonyl compounds to form acetylenic alcohols:

$$>CO + NaC{\equiv}CR \longrightarrow >C(OH)C{\equiv}CR$$

Acetylene also forms Grignard reagents by reaction with alkyl magnesium halides.

When acetylene is passed into an ammoniacal solution of cuprous chloride or silver nitrate, coprous acetylide Cu_2C_2(red) or silver acetyulide Ag_2C_2 (white) is precipitated. Both these compounds, when dry, explode when struck-or heated. Acetylenic copper compounds undergo coupling.

The acidic nature of hydrogen in acetylene is characteristic of hydrogen in the group ≡CH, and it has been suggested that this is because the C ≡ H bond has considerable ionic character due to resonance.

$$H{-}C \equiv C{-}H \longleftrightarrow H{-}C \equiv \overset{-\,+}{CH} \equiv C{-}H \longleftrightarrow \overset{+\,-}{HC} = \overset{-\,+}{CH}$$

There is, however, evidence to show that the electronegativity of a carbon atom depends on the number of bonds by which it is joined to its neighbouring carbon atom (Walsh, 1947). Since π-electrons are more weakly bound than σ-electron, the electron density round a carbon atom with π-bonds is less than that when only σ-bonds are present. Thus. a carbon atom having one π-bond has a slight positive charge compared with a carbon atom which has only s-bonds. Hence, the electronegativity of an sp^2 hybridised carbon atom is greater than that of an sp^3 hybridised carbon atom. Similarly, a carbon atom which has two π-bonds carries a small positive charge which is greater than that carried by a carbon atom with only one π-bond. Thus the electronagtivity of an *sp* hybridised carbon atom is greater than that of an sp^2 hybridised carbon atom. It therefore follows that the more s character a bond has, the more electronegative is that carbon atom. Thus the attraction for electrons by hybridised carbon will be sp > sp2 > sp3. Hence proton release (by heterolytic fission of the C—H bond) will occur mostreadily in acetylene, less readily in ethyulene, and least readily in methane. Alternatively, since the greater the electronegativity of an atom, the more readily it can accommodate a negative charge, stabilities of the following carbanions follow the order: $CH \equiv C^- > CH_2 = CH^{-1} > CH_3^-$

Acidities of hydrocarbons have been studied by determining the position of the following equilibrium in inert solvents:

$$R^1\ {}^-Na^+ + R_2H \rightleftharpoons R^1H + R^2—\ Na^+$$

The more the equilibrium is displaced to the right, the stronger is R_2H as an acid. The position of equilibrium may be determined in several ways, *e.g.*, (i) By passing carbon dioxide through the solution, only the sodium carboxylate of the more acidic hydrocarbon is formed. Since carboxylation upsets the equilibrium, this method actually gives information on which is the stronger acid (ii) By ultraviolet or visible spectroscopy; the anions usually differ from their parent hydrocarbons in these regions of the spectrum.

The structure of metallic carbides, *i.e.*, compounds formed between carbon and metals, is still a matter of dispute. It appears certain, so far, that the carbides of the strongly electropositive metals; N, K, Ca, Sr, Ba, are ionic —X-ray crystal analysis has shown the lattice of these carbides to be ionic, containing the ion $\overline{C} \equiv \overline{C}$. These carbides react with water to produce acetylene. Copper and silver carbids are not affected by water, and are explosive when dry. On account of these differences it seems likely that the carbides of these two metals are covalent, *e.g.*, Cu—C≡C—Cu. Thus these compounds may be regarded as *acetylides.*

In addition to the above carbides, there are a number of carbides which react with water or dilute acids to produce methane, *e.g.*, aluminium carbide; or a mixture of hydrocarbons, *e.g.*, uranium carbide which gives acetylene and other unsaturated hydrocarbons; iron carbide which gives methane and hydrogen.

May authors believe these carbides to be ionic, and include carbides of copper and silver in this group.

There is also one other group of carbides which are highly refractory, and which are extremely stable, chemcially; *e.g.*, vanadium carbide is not attacked by water or by drochloric acid even at 600° C. These carbides are believed to be *interstitial* compounds, *i.e.*, their lattice is not composed of ions, be resembles a metallic or atomic lattice.

Structure of acetylene : By reasoning similar to that used for ethylene, the structure of acetylene is show into be H—C°C—H, and may be represented as Fig. 2.4 (a). This representation however, appears to be inadequate. According to Coulson (1952), the two p-bonds from a charge cloud which has cylindrical symmetry about the carbon-carbon axis (Fig. 2.4b).

(a) (b)

Fig. 2.4

Homologues of Acetylene

Homologues of acetylene may be prepared by any of the following methods:

1. By the action of ethanolic potassium hydroxide on 1, 2-dialogen derivatives of the alkanes preparation of acetylene, method 2), *e.g.*, propyne from propene dibromide:

$$CH_3CHBrCH_2Br + 2KOH \xrightarrow{\text{ethanol}} CH_3C \equiv CH + 2KBr + 2H_2O$$

Sodamide in liquid ammonia usually gives better yields.

This method affords a simple means of introducing a triple bond into an organic compound, *e.g.*, p-butanol is catalytically dehydrated to but-1-ene, which on treatment with bromine gives 1, 2-dibromobutane, and this, when heated with ethanolic potassium hydroxide, yields but-1-yne;

$$CH_3CH_2CH_2CH_2OH \xrightarrow[350°C]{Al_2O_3} CH_3CH_2CH = CH_2 \xrightarrow{Br_2}$$

$$CH_3CH_2CHBrCH_2\,Br \xrightarrow[\text{ethanol}]{KOH} CH_3CH_2C \equiv CH$$

An interesting point in this connection is that open-chain 1, 2-diabilides give alkynes rather than conjugated denies on treatment with strong bases:

$$R^1CH = CHCH = CHR_2 \longleftarrow R^1CH_2CHXCHXCH_2R_2$$
$$\xrightarrow{-2HX} R_1CH_2C \equiv CCH_2R_2$$

2. Monosodium acetylide is treated with an alkyl halide, preferably a bromide, where upon an acetylene homologue is produced:

$$CH \equiv CH + Na \xrightarrow[NH_3]{liquid} CH \equiv CNa \xrightarrow{RX} CH \equiv CR + NaX$$

This reaction may be carried further as follows:

$$R^1C \equiv CH + Na \xrightarrow[NH_3]{liquid} R^1C \equiv CNa \xrightarrow{R^2X} R^1C \equiv CR_2 + NaX$$

In practice this method is limited to the use of primary alkyl halides, a since higher s- and t-halides give mainly alkenes when they react with the monosodium derivatives of acetylene or its homologues.

3. By the action of acetylene on a Grignard reagent (q.v.) and then treating magnesium complex with an alkyl halide:

$$CH \equiv CH + R\text{—}Mg\text{—}Br \longrightarrow RH + CH \equiv C\text{—}Mg\text{—}Mr$$
$$\xrightarrow{RBr} CH \equiv CR$$

The properties of the homologues of acetylene are very similar to those of acetylene, particularly when they are of they type RC ≡ CH, *i.e.*, contain the ≡CH group. Acetylene homologues isomerise when heated with ethanolic potassium hydroxide, the triple bond moving towards the centre of techcian, *e.g.*, but-1-yne isomerises to but-2-yne.

$$CH_3CH_2C \equiv CH \longrightarrow [CH_3CH = C = CH_2] \longrightarrow CH_3C \equiv CCH_3$$

There is a great deal of evidence to show that an allene is formed as an intermediate.

On the other hand, when alkynes are heated with sodamide in an inert solvent, *e.g.*, paraffin, the triple bond moves towards the end of the chain; *e.g.* but-2-yne gives the sodium derivative of but-1-yune, which is converted into but-1-yne by the action of dilute acid:

$$CH_3C \equiv CCH_3 + NaNH_2 \xrightarrow{paraffin} NH_3 + CH_3CH_2C \equiv CNa$$
$$\xrightarrow{H^+} CH_3CH_2C \equiv CH$$

This reaction affords a mean of stepping up the alkyne series by method 2 above.

Of the two alkynes, it is the 2-isomer which is the thermodynamically controlled product, but in the presence of the strong base sodamide, the formation of the *insoluble sodium salt* of the 1-isomer shifts the equilibrium in the direction of this product.

A very useful acetylene derivative for synthetic work is ethoxyacetylene. This may conveniently be prepared by the action of sodamide on chloroacetladehyde diethyl acetal (Jones et al., 1954).

$$ClCH_2\ CH\ (OC_2H_5)_2 \xrightarrow{NaNH_2} CH \equiv COC_2H_5$$

Lithium derivatives of alkynes can be prepared by interaction of the alkyne and lithium amide in dioxane (Schlubach et al., 1958).

Cuprous acetyides undergo oxidative coupling when heated in acetic acid in the presence of air (Weedon et. al, 1958).

$$2RC \equiv CCU \xrightarrow{O_2} RC \equiv C—C \equiv CR$$

Eglington et al. (1959) have used cupric acetate in pyridine instead of oxygen (see p. 561).

A variation of the method has been introduced by Chodkiewicz (1957) who used a 1-alkyne and a bromo derivative of a 1-alkyne (prepared by the action of sodium hybromite on a 1-alkyne. This method has also been extended to 'crossed' oxidative coupling *e.g.*,

$$MeCH \equiv CMeC \equiv CBr + CH \equiv C\ (CH_2)_8\ CO_2H$$

$$\xrightarrow{Ca^+} MeCH = CMeC \equiv C—C \equiv C(CH_2)_8\ CO_2H$$

EXERCISES

1. What alkenes could you expect from the following compound? Give the experimental conditions:
   (i) Me_3CCl;
   (ii) $Me_2C(OH)CMe_3$;
   (iii) Me_2CHCH_2OAe;
   (iv) $MeCH_2CH(OAC)Me$.
2. Draw only the geometrical isomeric forms of; (i) n-pentenes; (ii) isopentenes; (iii) monobromopropene.

3. Draw the energy profiles for the reactions:

   (i) $MeCH = CH_2 + HX \longrightarrow MeCH_2CH_2^+ + X^-$

   (ii) $MeCH = CH_2 + HX \longrightarrow Me\overset{+}{C}HMe + X^-$

   Give reasons for your answers.
4. Write out the structures (ignoring stereochemistry) of the isomeric pentenes, and name them by the IUPAC system. Given the structures of the products formed from each onozonolysis.
5. Using bond energies, estimate ΔH for the following reactions:

   (i) $MeCH = CH_2 + HBr \longrightarrow MeCH_2CH_2Br$

   (iI) $MeCH = CH_2 + HBr \longrightarrow MeCHCHBrmE$

   What would you predict from the results? Are these predictions correct? If not, why not? Under what experimental conditions are the *equations* correct?
6. Addition of HCl to 3.3-dimethyl but-1-ene gives two isomers alkyl chloridies Explain.
7. Arrange the following carbonium ions in order of increasing stability, and give your reasons.

   $Me\overset{+}{C}HMe$ $\quad CH_2 = CHCH_2^+ \quad MeCH_2CH_2^+$
8. Can you compare the stabilities of pent-1-ene, cis- and trans-pent-2-enes and 3-methylbut-1-ene.
9. Using the Boord synthesis, how would you prepare:

   (i) but-2-ene;

   (ii) hex-3-ene?
10. Discuss the mechanisms of the following reactions:

    (i) $MeCH = CH_2 + HOBr \longrightarrow MeCHOHCH_2Br$

    (ii) $CCl_3CH = CH_2 + HOBRr \longrightarrow CCl_3CHBrCH_2OH$
11. Compound A (C, 88.9%; H, 11.1%) M.W. 54 on catalytic hydrogenation, gives B (C, 82.8%; H,17.2). The i.r. spectrum of A shows a band in the region 2240 cm^{-1}. What is A?
12. Draw the energy profile for the addition of HI to propene. Explain your diagram.

13. Draw the energy profiles for : (i) the hydrogenation of ethylene with and without acatalyst; (ii) the catalysis hydrogenation of *cis*- and *trans*- but-2-enes.

14. Isobutene, in the presence of H_2SO_3, forms a mixture of two isomeric alkenes, C_8H_{16}. Formulate the reaction. How could you distinguish between the isomers chemically?

15. Formulate the reactions between but-1-ene, in the presence of a small amount of benzoyul peroxide.

    (i) CCl_4; (ii) $CBrCl_3$. Give your reasons.

16. How many NMR signals would you expect for the following compounds? If a signal is other than a indicate this by (m):

    (i) 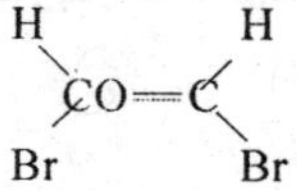 (H)(Br)CO=C(H)(Br)

    (ii) 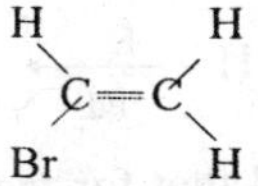 (H)(Br)C=C(H)(H)

    (iii) $CH \equiv CCH_2Br$

    (iv) $CH_2 = C = CH_2$

    (v) $CH_3C \equiv CH$

    (vi) 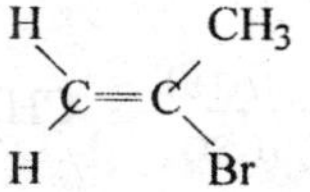 (H)(H)C=C(CH_3)(Br)

17. Some of the mass spectra peaks given by the three methylpen-1-enes. A, B and C are listed in the table. What are A, B and C? Give reasons for your answer.

**	27	28	29	41	42	43	55	56	57	69	84
A	48	7	40	79	21	75	100	39	—	71	30
B	35	8	28	80	29	8	43	100	44	36	41
C	33	4	6	72	33	100	8	44	20	13	11

18. Complete the following equations, and comment where necessary:

    (i) $MeCH_2CH_2CMe = CHMe \xrightarrow{SeO_2} ?$

    (ii) $MeCH_2C \equiv CBr + CH \equiv CMe \xrightarrow{Cu^+} ?$

(iii) $Me_2CHCH = CH_2 + HBr \longrightarrow ?$

(iv) $MeCH_2C \equiv CH \xrightarrow{Na/NH_3} ?$

(v) $BrCH_2CH_2Br \xrightarrow{?} C_2H_2$.

(vi) $MeCH_2CH_2OH \xrightarrow{?} MeCH = CH_2$

(vii) $MeCHBrCH_2Br \xrightarrow[H_2SO_4]{Hg^{2+}} ?$

(viii) (cyclohexane ring)$-C{\equiv}CH \xrightarrow[Hg_2SO_4]{Hg^2}$

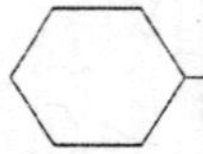

19. What are the geometries of:

(i) but-1-yne; (ii) but-2-yne; (iii) hept-2-en-5-yne?

20. Complete the following equation:

$$CH_3COCH_3 \xrightarrow{NaNH_2} ? \xrightarrow{C_2H_2} ?$$

$$\xrightarrow{H^+} (CH_3)_2C(OH)C \equiv CH \xrightarrow[Pd]{H^+} ? \xrightarrow[400^\circ C]{Al_2O_3}$$

21. For cis-but-2-ene is 65.86 kJmol^{-1} and that for *trans*-but-2-ene is 62.97 kJ mol^{-1}. What is the value of K at 25° C for the equilibrium mixture? Does either isomer form this equilibrium mixture at 25° C? Discuss. Explain the course of the following reaction:

$CH_2 = CHCHOHCH = CHMe \xrightarrow[H_2SO_4]{MeOH} CH_2 = CHCH=$ CHCH(OMe) Me

22. Compound A (C, 87.1%; H, 12.8%), on catalytic reduction, gives B (C, 84.1%; H, 15.9%). Ozonolysis of gives acetic acid, acetone and pyruvic acid ($MeCOCO_2H$). What are A and B?

3

HALOGEN DERIVATIVES OF THE ALKANES

Halogen derivatives of the alkanes are divided into mono, di-, tri, etc, substitution products according to the number of halogen a toms in the molecule.

ALKYL HALIDES

The alkyl halides have the general formula $C_nH_{2n+1}X$ or R.X, where X denotes chlorine bromine or iodine. Fluorine is not included, since fluorides do not behave like the other halides.

GENERAL METHODS OF PREPARATION

1. The method most widely used is the replacement of the hydroxyl group of an alcohol by an X atom. Halogen acid is the most common reagent used with alcohols, but the conditions depend on both the structure of the alcohol and the particular acid.

$$ROH + HX \xrightarrow{\text{heat}} RX + H_2O$$

Hydrogen chloride is used with zinc chloride (*Groves' process*) for primary and secondary alcohols, but tertiary alcohols readily react with concentrated hydrochloric acid in the absence of zinc chloride.

Zinc chloride as a Lewis acid and consequently can co-ordinate with the alcohol.

$$ROH + ZnCl_2 \longrightarrow R\!-\!\overset{+}{\underset{\substack{\vdots \\ H}}{O}}\!-\!\overset{-}{Z}nCl_2 \xrightarrow{S_N1} R+ + [HO\!-\!ZnCl_2]$$

The R—O bond is weakened and so the complex readily forms a carbonium ion (S_N1) mechanism. If the nature of R^+ is such that it can undergo rearrangement, the product will be a mixture of isomeric alkyl chlorides. When the concentration of zinc chloride is low, the reaction is still catalysed, but no rearrangement occurs (S_N2 mechanism).

$$ZnCl_2 + HCl + ROH \longrightarrow$$

$$ZnCl_3^- + ROH_2^+ \xrightarrow{S_N2} ZnCl + RCl + H_2O$$

Fridine and dimethylamine also catalyse the reaction between alcohols and hydrochloric acid without rearrangement.

The formation of the various halides by rearrangement is one example of 1, 2-shifts. In this type of rearrangement a group migrates from a *carbon* atom (the *migration origin)* to an *adjacent* atom (the *migration terminus(* which is usually carbon or nitrogen. The rearrangement may be expressed in the following general form.

$$\underset{|}{\overset{Z}{}}A—B—Y \longrightarrow \bar{Y} + \overset{Z}{A}—B^+ \longrightarrow \overset{+}{A}—\overset{Z}{B} \xrightarrow[\text{rageet}]{\text{nucleopbilic}} \text{products}$$

When Y ionises, B is left with only six valency electrons (open sextet), and consequently carries a positive charge Z migrates *with its bonding pair of electrons,* and A now has an open sexted Attack now takes place at A by Y^- or some other nucleophilic reagent to form the products. Then mechanism was first suggested by Whit more (1932) and is often referred to as the *Whit more mechanism.*

The nature of Z is quite varied, the 'key' atom (*i.e.*, the atom joined to the bonding pair which migrates) may be carbon, halogen, oxygen, etc., If Z has a lone pair of electrons, *e.g.*, halogen, then this pair may be used in the migration to give a cyclic intermediate, *i.e.*, a bridged or non chemical carbonium ion.

$$\overset{Z}{A}—B^+ \longrightarrow A\overset{\overset{+}{Z}}{—}B \longrightarrow \text{product}$$

If Z is an aryl group, then this group can supply an electron pair to form a bridged ion. If Z is an alkyl group, then there are o lone pairs or multiple bonds, but again it is believed that a bridged ion is possible, *i.e.*,

$$—\overset{R}{C}—C^+ \longrightarrow C\overset{R}{=\!=\!=}C$$

This type of bridged ion contains three *partial bonds formed form one pair of electrons. The* formation of these bridged ions in 1,2-shifts is an example of *nighbouring group participation,* and when the rate is increased because of this effect, the rearrangement is said to be *an chemically assisted* (Winstein *et al.*, 1953).

A very important point about 1,2-shifts is that they are *intramolecular*', *i.e.*, Z is never actually free. If the mechanism were such that Z was free during is migration, then the rearrangement would be called *intermolecular.* The intramolecular nature of the 1,2-shifts has been established by using a group Z which contains an asymmetric carbon atom attached to A, i.e, have type of C—A. If Z actually separated during the migration, then it means that the free migrating that is carbanion, *i.e.* abC:⁻. Examination of such carbanions has shown that none ever mains its configuration, it recemises, *i.e.*, half the molecules have their configuration inverted, and so the product is no longer optically active. Thus if Z moves with its bonding pair and the configuration is *retained* in the product, than Z can never been free, *i.e.*, the rearrangement is intramolecular.

We can now illustrate the foregoing discussion with neopentyl alcohol as our example. Whit more *et al.* (1932) showed that the reactions undergone by neopentyl alcohol and neopentyl halides are of two types. (i) replacement reactions, which occur very slowly (if at all) and produce neopentyl compounds, (ii) replacement (and elimination) reactions, which occur very fast and produce t-amyl compounds.

Hughes and Ingold (1996) have shown that the slow reactions take place by the S_N2 mechanism and consequently without rearrangement, *e.g.*, neopentyl bromide reacts with ethanolic sodium ethoxide to give ethyl neopentyl ether.

$$\underset{BrCH_2}{\overset{Me_3C}{|}} + OEt^- \rightleftharpoons \underset{\overset{\delta-}{Br}\cdots CH_2\cdots \overset{\delta-}{OEt}}{\overset{Me_3C}{|}} \xrightleftharpoons{Br^-} \underset{CH_2OEt}{\overset{Me_3C}{|}}$$

On the other hand when neopentyl alcohol reacts with anhydrous hydrogen bromide, the main product is t-amyl bromide (72 per cent), together with primary and secondary bromides. The arranged bromides are the result of 1,2-shifts (methyl and hydride)

$$Me_3CCH_2OH \xrightleftharpoons{HBr^+} Me_3CCH_2—\overset{+}{O}H_2 \rightleftharpoons Me_3C—\overset{+}{C}H_2 + H_2O$$

$$Me_3C—\overset{+}{C}H_2 \xrightarrow{Br^-} Me_3CCH_2Br$$

$$Me_3C—\overset{+}{C}H_2 \xrightarrow{\text{Me shift}} Me_2\overset{+}{C}CH_2Me \xrightarrow{Br^-} Me_2CBrCH_2Me$$

$$Me_2\overset{+}{C}CH_2Me \xrightarrow{\text{H shift}} Me_2CH\overset{+}{C}HMe \xrightarrow{Br^-} Me_2CHCHBrMe$$

The 'deriving force' of the rearrangement is probably due to the stabilities of carbonium being tertiary > secondary > primary.

Any alkyl halide may be prepared by the action of a phosphorus halide on the alcohol Phosphorus pentachloride gives variable yields depending on the alcohol.

$$ROH + PCl_5 \longrightarrow RCl + HCl + POCl_3$$

Straight- chain primary alcohols react with phosphorus trihalides to give unrearranged halides but many secondary alcohols undergo rearrangement.

$$PX_3 + 3ROH \longrightarrow (RO)_3P + 3HX \longrightarrow (RO)_3$$

$$PH \xrightarrow[S_N2]{XH} RX + (RO)_2P(H)O \xrightarrow[S_N2]{HX}$$

$$RX + (RO)P(H)(OH)O \xrightarrow[S_N2]{HX} RX + H_3PO_3$$

Thionyl chloride reacts with straight-chain primary alcohols without rearrangement in the presence or absence of pyridine (Darzens procedure. 1911), the reaction proceeds via a chlorosulphite.

$$ROH + SOCl_2 \longrightarrow HCl + ROSOCl$$

$$\longrightarrow Cl^- + ROSO^+ \xrightarrow{S_N2} RCl + SO_2$$

2. By direct halogenation *i.e.*, substitution reactions of the alkanes with halogen, and this ml be brought about by light, catalysts or heat.

Photohalogenation is carried out by treating the alkane with chlorine (or bromine) at ordinate temperature in the presence of light. The reaction takes place by a free- radical chain mechanism the initiation being brought about by the formation of chlorine atoms by absorption of lights

$$Cl_2 \xrightarrow{hv} 2Cl^{\cdot}$$

$$Cl. + H—CH_3 \longrightarrow Cl...H...CH_3 \longrightarrow HCl + CH_3$$

$$CH_3. + Cl—Cl \longrightarrow CH_3...Cl...Cl \longrightarrow CH_3Cl + Cl.$$

$$Cl. + H—CH_2Cl \longrightarrow Cl...H...CH_2Cl \longrightarrow HCl + CH_2Cl.$$

$$ClCH_2^{\cdot} + Cl—Cl \longrightarrow ClCH_2...Cl...Cl \longrightarrow CH_2Cl_2 + Cl. \text{ etc.}$$

The termination of the chain reaction may take place by adsorption of the chlorine atom on the walls of the containing vessel, of by two chlorine atoms combining with each other form a chlorine molecule.

There appears to be no evidence to show that CH_3, CH_2Cl radicals combine to terminate the chain reaction, but another termination step is the combination of R, with Cl. (to give RCl).

Thermal halogenation. Thermal chlorination has been studied in great detail by Hass. Mass and their co-workers (1935 onwards), and as a result of their work they suggested a number rules for chlorination.

GENERAL METHODS OF PREPARATION AND PHYSICAL PROPERTIES OF HALOGEN DERIVATIVES OF HYDROCARBONS

general Methods of Preparation of Haloalkanes and Haloarenes

General methods of preparation of haloalkanes and haloarenes are described below.

From Hydrocarbons

Halo derivatives of hydrocarbons can be prepared by various methods. Some of these methods are described below.

By free radical halogenation : Direct chlorination or bromination of alkanes in the presence of heat or ultraviolet radiation proceed via a free radical mechanism. Usually a mixture of isomeric mono- and poly haloalkanes is obtained. Therefore, halogenation of alkanes by free radical mechanism is not suitable for obtaining a pure haloalkane.

$$\underset{\text{bu tan e}}{CH_3CH_2\,CH_2CH_2} + Cl_2 \xrightarrow{\text{UV / heat}}$$

$$\underset{\text{isomeric monochlorobu tan es}}{CH_3CH_2CH_2CH_2Cl,\ CH_3CH_2CH(Cl)CH_3}$$

$$\left.\begin{array}{ll} CH_3CH(Cl)CH(Cl)CH_3, & CH_3CH_2CH(Cl)CH_2Cl \\ CH_3CH(Cl)CH_2CH_2Cl\ , & CH_3CH_2CH_2CHCl_2 \\ CH_3CH_2CCl_2CH_3 \quad , & ClCH_2CH_2CH_2CH_2Cl \\ \multicolumn{2}{c}{\text{isomeric dichlorobu tan es}} \end{array}\right] + \underset{\text{derivatives}}{\text{Higher}}$$

The direct halogenation method, however is useful in certain cases. For example,

(i) When no isomeric product is formed and the mixture can be separated into different products by fractional distillation due to difference in the boiling points of the products formed

$$\underset{\text{methane}}{CH_4} \xrightarrow[-\,HCl]{+\,Cl_2,\,h\nu} \underset{\text{chloromethane}}{CH_3Cl} \xrightarrow[-\,HCl]{+\,Cl_2} \underset{\text{dichloromethane}}{CH_2Cl_2}$$

$$\xrightarrow[-\,HCl]{+\,Cl_2} \underset{\text{trichloromethane}}{CHCl_3} \xrightarrow[-\,HCl]{+\,Cl_2} \underset{\substack{\text{tetrachloromethane}\\ \text{(carbon tetrachloride)}}}{CCl_4}$$

$$\underset{\substack{\text{2,2-dimethylpropane}\\ (\textit{neo}\text{-pentane})}}{CH_3-C(CH_3)_2-CH_3} + Cl_2 \xrightarrow{\text{heat/UV}} \underset{\substack{\text{1-chloro-2,2-dimethylpropane}\\ (\textit{neo}\text{-pentyl chloride})}}{CH_3-C(CH_3)_2-CH_2Cl} + HCl$$

In halogenation reactions,

(a) Chlorine reacts with alkanes readily, but the reaction with bromine is solver. The reaction with iodine is reversible. Thus, the reactivity of halogens follow the order.

$$Cl_2 > Br_2 > I_2$$

Reaction with fluorine is highly exothermic and violent. It leads to the formation of polyfluoro compound and cleavage of the carbon-carbon bonds.

(b) During halogenation reactions, the tendency of the replacement of hydrogen atom of an alkane by a halogen is

tertiary alkane > secondary alkane > primary alkane

(c) By this method, generally a mixture of haloalkane and polyhaloalkanes is obtained. These can be separated by fractional distillation.

(ii) When pure compounds are not required, such as those used as industrial solvents. The mixture of different products can be enriched in the desired product either by fractional distillation or by adjusting the ration of the reactants. For example, when methane is chlorinated in the presence of excess of methane, the major product formed is chloromethane

$$\underset{\text{excess}}{CH_4(g)} + Cl_2(g) \xrightarrow{\text{heat/UV}} \underset{\text{chloromethane}}{CH_3Cl} + HCl$$

The unreacted methane is separated and recycled.

Chloromethane and bromomethane are obtained industrially by this method.

(iii) When difference in the reactivity of different types of hydrogens in the hydrocarbons molecule is so large that only one product is formed under suitable conditions.

The order of reactivity for different types of hydrogens is

$$\underbrace{\text{Benzylic} \approx \text{Allyic}}_{\text{high reactivity}} > \text{Tertiary} > \text{Secondary} > \text{Primary} > \underbrace{\text{Vinylic} \approx \text{Aryl}}_{\text{these hydrogens do not usually participate in free radical halogenation reactions}}$$

Thus *benzylic and allylic hydrogens can be replaced by halogens without any side reaction.* For example,

$$\underset{\text{propane}}{CH_2=CH-CH_3} + Cl_2 \xrightarrow[\text{gas phase}]{500\text{–}600°C} \underset{\text{3-chloropropene}}{CH_2=CH-CH_2Cl_2} + HCl$$

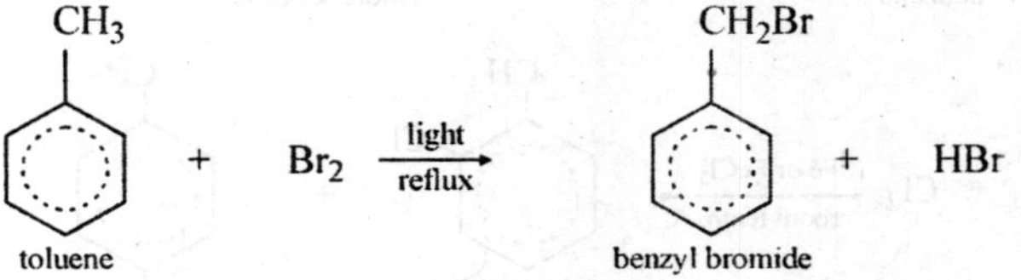

$$\underset{\text{toluene}}{C_6H_5CH_3} + Br_2 \xrightarrow[\text{reflux}]{\text{light}} \underset{\text{benzyl bromide}}{C_6H_5CH_2Br} + HBr$$

Fig. 3.1

The high reactivity of allylic and benzylic hydrogens is due to the stabilization of the allylic and benzylic free radicals formed during the reactions by resonance.

$$CH_2=CH-\dot{C}H_2 \underset{\text{resonance in allyl radical}}{\longleftrightarrow} \dot{C}H_2-CH=CH_2$$

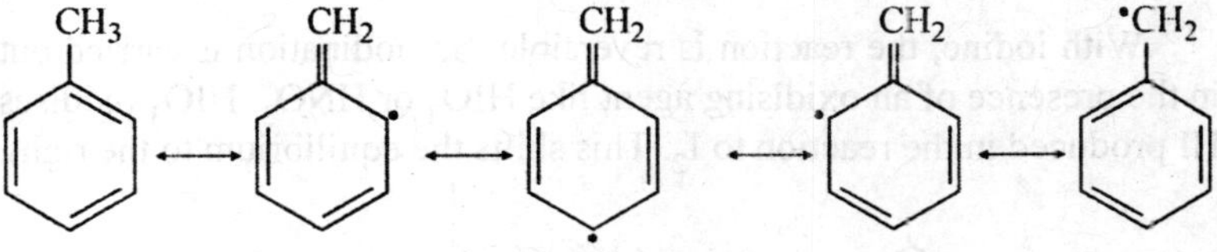

resonance in benzyl radical

Fig. 3.2

1. *By electrophilic substitution :* The chloro and bromo derivatives of arenes can be obtained by electrophilic substitutions of arenes with chlorine or bromine in the presence of iron or iron (III) halide or any Lewis acid as catalyst.

Low temperature, and presence of a halogen carrier favour nuclear substitution. Chlorination and bromination may be very conveniently carried out at room temperature in the presence of a Lewis acid such as, $FeCl_3$, FeBr, $AlCl_3$, etc. The Lewis acid acts as a *halogen carrier.*

In actual practice, chlorination and bromination are done in the presence of iron. Iron gets converted into the corresponding halide by the halogen being used.

$$\underset{\text{arene}}{ArH} + \underset{(X = Cl,\ Br)}{X_2} \xrightarrow{\text{Lewis acid}} \underset{\text{haloarene}}{ArX} + HX$$

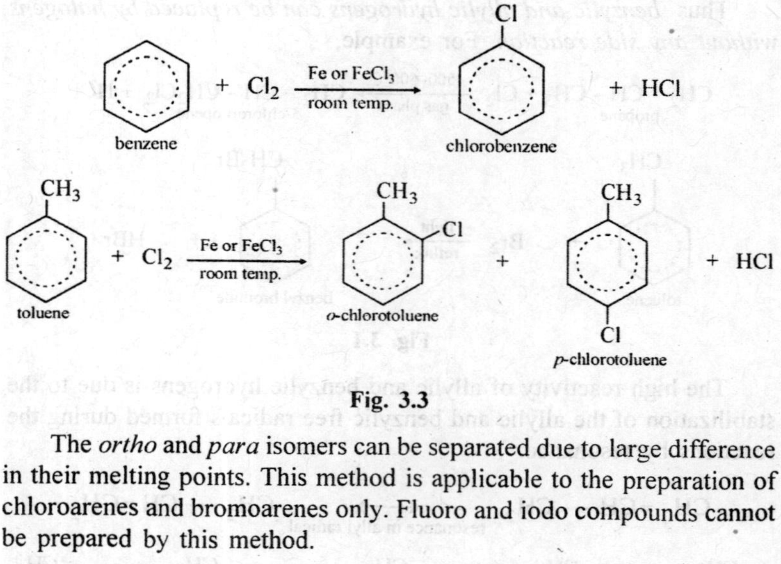

$$\underset{\text{benzene}}{C_6H_6} + Cl_2 \xrightarrow[\text{room temp.}]{\text{Fe or FeCl}_3} \underset{\text{chlorobenzene}}{C_6H_5Cl} + HCl$$

$$\underset{\text{toluene}}{C_6H_5CH_3} + Cl_2 \xrightarrow[\text{room temp.}]{\text{Fe or FeCl}_3} \underset{o\text{-chlorotoluene}}{o\text{-}ClC_6H_4CH_3} + \underset{p\text{-chlorotoluene}}{p\text{-}ClC_6H_4CH_3} + HCl$$

Fig. 3.3

The *ortho* and *para* isomers can be separated due to large difference in their melting points. This method is applicable to the preparation of chloroarenes and bromoarenes only. Fluoro and iodo compounds cannot be prepared by this method.

With iodine, the reaction is reversible. So, iodination is carried out in the presence of an oxidising agent like HIO_3 or HNO_3, HIO_3 oxidises HI produced in the reaction to I_2. This shifts the equilibrium to the right.

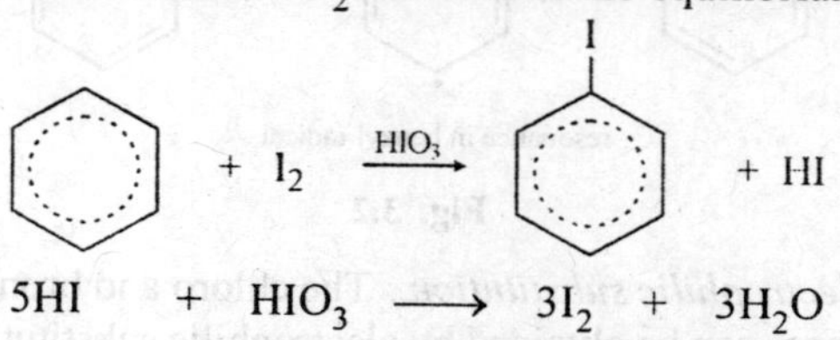

$$C_6H_6 + I_2 \xrightarrow{HIO_3} C_6H_5I + HI$$

$$5HI + HIO_3 \longrightarrow 3I_2 + 3H_2O$$

Fig. 3.4

2. *By electrophilic addition :* Alkenes and alkynes undergo electrophilic addition of halogens (Cl_2, Br_2) and hydrogen halides (HCl, HBr, HI) to form halogen derivatives.

$$\underset{\substack{\text{ethene} \\ \text{(ethylene)}}}{CH_2{=}CH_2} + HX \longrightarrow \underset{\substack{\text{haloethane} \\ \text{(ethyl halide)}}}{CH_3CH_2X}$$

$$\underset{\text{but-2-ene}}{CH_3-CH{=}CH-CH_3} + HX \longrightarrow \underset{\text{2-halobutane}}{CH_3-\underset{H}{\underset{|}{C}}H-\underset{X}{\underset{|}{C}}H-CH_3}$$

$$R-C{\equiv}CH + HX \longrightarrow \underset{\text{monohalo}}{R-\underset{X}{\underset{|}{C}}{=}\underset{H}{\underset{|}{C}}H} \xrightarrow{+HX} \underset{\text{dihalo}}{R-\underset{X}{\underset{|}{\overset{X}{\overset{|}{C}}}}-\underset{H}{\underset{|}{\overset{H}{\overset{|}{C}}}}-H}$$

In symmetrical alkenes, the carbon atoms across the double bond are equivalent. So, the addition of an acid molecule to symmetrical alkenes gives only one product.

The addition of halogen acid to unsymmetrical alkenes proceeds in accordance with the *Markownikoff's rule,* and secondary and tertiary haloalkanes are obtained.

Markownikoff's rule : According to this rule, the *negative part of the molecule to be added across a double or a triple bond gets attached to the carbon atom containing lesser number of hydrogen atoms.*

$$\underset{\text{unsymmetrical alkene}}{R-CH{=}CH_2} + HX \longrightarrow \underset{\text{2-haloalkane}}{R-\underset{X}{\underset{|}{C}}H-\underset{H}{\underset{|}{C}}H_2} \text{ or } R-\underset{X}{\underset{|}{C}}H-CH_3$$

$$\underset{\text{propene}}{CH_3CH{=}CH_2} + HBr \longrightarrow \underset{\text{2-bromopropane}}{CH_3-\underset{Br}{\underset{|}{C}}H-\underset{H}{\underset{|}{C}}H_2} \text{ or } CH_3-\underset{Br}{\underset{|}{C}}H-CH_3$$

Peroxide effect : The addition to a double bond in unsymmetrical haloalkanes takes place in accordance with the Markownikoff's rule. However, in the presence of peroxides *e.g.*, benzoyl peroxide (C_6H_5 — CO — O — O — CO — C_6H_5), the addition of hydrogen halide to an unsymmetrical alkene takes place contrary to the Markownikoff's rule. This is called *peroxide effect, or Kharasch effect.*

$$\underset{\text{propene}}{CH_3 - CH = CH_2} + HBr \xrightarrow{\text{Peroxide}} \underset{\text{1-bromopropane}}{CH_3 - CH_2 - CH_2Br}$$

As we see, in the presence of peroxide, the negative part of the molecule to be added goes to the carbon atom containing larger number of hydrogen atoms. This is also called *anti-Markownikoff's rule.*

This is because in the presence of a peroxide, the addition takes place by a free-radical mechanism.

FROM ALCOHOLS

This is the most widely used method for the preparation of haloalkanes. The hydroxyl group of alcohols may be replaced by halogen atom by using hydrogen halides (HX), phosphorus halides (PCl_5, PCl_3) or thionyl halides, ($SOCl_2$).

1. *By the action of hydrogen halide (halogen acids)* : Action of a hydrogen halide on an alcohol in the presence of a dehydrating agent such as, anhydrous zinc chloride, *(Groove's process),* or conc., sulphuric acid, gives haloalkanes. Hydrobromic acid may be generated in the reaction mixture (*in situ)* by the reaction of NaBr or KBr with conc. H_2SO_4.

$$\underset{\text{alcohol}}{ROH} + HX(g) \xrightarrow{ZnCl_2 \text{ anhyd}} \underset{\text{haloalkane}}{R.X} + H_2O$$

$$ROH + NaBr + H_2SO_4 \xrightarrow{\Delta} RBr + NaHSO_4 + H_2O$$

Reactivites of acids follow the order.

HI > HBr > HCl

Reactivites of alcohols follows the order.

Tertiary > Secondary > Primary

$$\underset{\text{ethyl alcohol}}{C_2H_5OH} + HCl(g) \xrightarrow{ZnCl_2 \text{ anhyd.}} \underset{\text{chloroethane}}{C_2H_5Cl} + H_2O$$

$$C_2H_5OH + HBr(g) \xrightarrow{\Delta, H_2SO_4} \underset{\text{bromoethane}}{C_2H_5Br} + H_2O$$

$$C_2H_5OH + HI(g) \xrightarrow{\Delta} \underset{\text{iodoethane}}{C_2H_5I} + H_2O$$

$$\underset{\text{propan-1-ol}}{CH_3 - CH_2 - CH_2OH} + HCl(g) \xrightarrow{ZnCl_2 \text{ anhyd}}$$

$$CH_3-CH_2-CH_2Cl+H_2O$$
1–chloropropane

$$\underset{\text{propane-2-ol}}{CH_3-\underset{|}{\overset{}{C}}\underset{OH}{}-CH_3} + HCl(g) \xrightarrow{ZnCl_2\ anhyd.} \underset{\text{2-chloropropane}}{CH_3-\underset{Cl}{\underset{|}{CH}}-CH_3} + H_2O$$

Tertiary alcohols are very reactive. So, tertiary alcohols react with conc. HCl even in the absence of zinc chloride.

$$\underset{\text{propane-2-ol}}{CH_3-\overset{CH_3}{\underset{CH_3}{C}}-OH} + \text{con. HCl} \longrightarrow \underset{\text{2-chloro-2-methylpropane (tert-butyl chloride)}}{CH_3-\overset{CH_3}{\underset{CH_3}{C}}-Cl} + H_2O$$

2. *By the action of thionyl chloride :* When thionyl chloride (SOCl2) and alcohol are heater in the presence of pyridine, haloalkane is obtained.

This method is preferred beacause the other products of the reaction (HCl and SO2) being escape out leaving behind pure haloalkane.

$$ROH+SOCl_2 \xrightarrow{\text{pyridine}} RCl + HCl(g)+SO_2(g)$$

Example:

$$CH_3CH_2OH+SOCl_2 \xrightarrow{\text{pyridine}} C_2H_5Cl+HCl(g)+SO_2(g)$$

Haloarenes cannot be prepared by similar reaction using phenols. This is because due to resonance, the carbon—oxygen bond in phenols develops a double bond character and is therefore, stronger than the carbon—oxygen single bond in alcohols. As a result, the carbon-oxygen bond in phenols is not cleaved easily.

:ÖH I ⟷ :ÖH II ⟷ $\overset{\oplus}{:\text{OH}}$ III ⟷ $\overset{\oplus}{:\text{OH}}$ IV ⟷ $\overset{\oplus}{:\text{OH}}$ V

Resonance structures of phenol

Fig. 3.5

3. *By the action of phosphorus halides :* Alcohols on refluxing with phosphorus penta or trihalides give haloalkanes. For preparing chloroalkanes, PCl_5 or PCl_3 is used. To prepare a bromoalkane, or an iodoalkane, the alcohol is heated with red phosphorus and bromine ($P + Br_2$) or red phosphorus and iodine ($P + I_2$) respectively.

$$\underset{\text{alcohol}}{ROH} + PCl_5 \xrightarrow{\Delta} \underset{\text{chloroalkane}}{RCl} + POCl_3 + HCl$$

$$\underset{\text{alcohol}}{3ROH} + PCl_3 \xrightarrow{\Delta} \underset{\text{chloroalkane}}{3RCl} + H_3PO_3$$

Example:

$$\underset{\text{propanol}}{CH_3CH_2CH_2OH} + PCl_3 \xrightarrow{\Delta} \underset{\text{1-chloropropane}}{CH_3CH_2CH_2Cl} + H_3PO_3$$

$$\underset{\text{propan-1-ol}}{CH_3CH_2CH_2OH} + PCl_5 \xrightarrow{\Delta} \underset{\text{1-chloropropane}}{CH_3CH_2CH_2Cl} + POCl_3 + HCl$$

$$\underset{\substack{\text{propane-2-ol} \\ \text{(secondary alcohol)}}}{CH_3-\underset{\substack{|\\ OH}}{CH}-CH_3} + PCl_3 \longrightarrow \underset{\text{2-chloropropane}}{CH_3-\underset{\substack{|\\ Cl}}{CH}-CH_3} + POCl_3 + HCl$$

$$\underset{\text{ethanol}}{6CH_3-CH_2OH} \xrightarrow{2P + 2Br_2} \underset{\text{bromoethane}}{6CH_3-CH_2Br} + 2H_3PO_3$$

$$\underset{\text{ethanol}}{6CH_3-CH_2OH} \xrightarrow{2P + 3I_2} \underset{\text{iodoethane}}{6CH_3-CH_2I} + 2H_3PO_3$$

$$\underset{\text{propan-1-ol}}{6CH_3-CH_2-CH_2OH} \xrightarrow{2p + 3I_2} \underset{\text{1-iodopropane}}{6CH_3-CH_2-CH_2I} + 2H_3PO_3$$

1. *Preparation of iodoalkanes :* An iodoalkane is often prepared from the corresponding bromo or chloro alkane by treating it with a solution of sodium iodide in acetone.

   The chlorine or bromine atoms of the haloalkane is replaced by iodine atoms in this reaction (*Finkelstein reaction*). This reaction gives the best results with primary halides.

$$\underset{(X = Cl, Br}{R-X} + NaI \xrightarrow{\text{acetone, reflux}} R-I + NaX(s)$$

$$\underset{\text{chloroethane}}{CH_3CH_2Cl} + NaI\ (\text{in acetone}) \xrightarrow{\text{reflux}} \underset{\text{iodoethane}}{CH_3CH_2I} + NaCl(s)$$

NaBr, and NaCl are less soluble in acetone than NaI. So they precipitate from the solution and thus preventing the backward reaction to occur.

2. *Preparation of fluoro derivatives of alkanes :* Fluoro derivatives of alkanes can be obtained by heating a chloroalkane or bromoalkane with a metallic fluoride such as AgF, Hg_2F_2, CoF_3 or SbF_3, (*Swarts reaction*). For example,

$$\underset{\text{bromomethane}}{CH_3Br} + AgF \xrightarrow{\text{heat}} \underset{\text{fluoromethane}}{CH_3F} + AgBr$$

For di or tri halo derivatives with halogen atoms on the same carbon atom (in the alkane) commonly SbF_3 or CoF_3 is used

$$\underset{\text{2,2-dichloropropane}}{3CH_3CCl_2CH_3} + \underset{\text{antimony trifluoride}}{2SbF_3} \xrightarrow{\text{heat}} \underset{\text{2,2-difluoropropane}}{3CH_3CF_2CH_3}$$

3. *Preparation of Haloarenes From Diazonium Salts*

Diazonium salts are wonderful strating materials for many aromatic compounds. These will be taught to you in class XII. Here we provide a brief account of their use in preparative chemistry of haloarenes. Haloarenes can be very conveniently prepared from diazonium salts. Diazonium salts are very unstable compounds. Therefore, freshly, prepared solution of a diazonium salt is used for these preparations.

A diazonium salt is obtained by treating an aromatic primary amine with nitrous acid at temperature below 5°C. This reaction is known as *diazotisation reaction.*

In actual practice, an ice-cold aqueous solution of $NaNO_2$ (sodium nitrite) is slowly added into an ice-cold solution of aniline in HCl (maintained at < 5°C). The reactions taking place are.

$$NaNO_2 + HCl \xrightarrow{< 5°C} HNO_2\ (aq) + NaCl$$

$$\underset{\text{aniline}}{C_2H_5 - NH_2} + HNO_2\ (aq) + HCl(aq) \xrightarrow{< 5°C}$$

$$\underset{\text{benzenediazonium chloride}}{C_6H_5 - N^+ \equiv NCl^-} + 2H_2O$$

The ice-cold solution of benzendiazonium salt is used as such.

4. *From Chloroalkanes, Bromoalkanes By Halogen Exchange Method*

Fluro and iodoalkanes cannot be prepared by direct fluorination and iodination. For preparing fluoro and iodo derivatives of alkanes, generally halogen exchange method is used. Such methods are described below.

Preparation of different haloarenes from diazonium salts is described below.

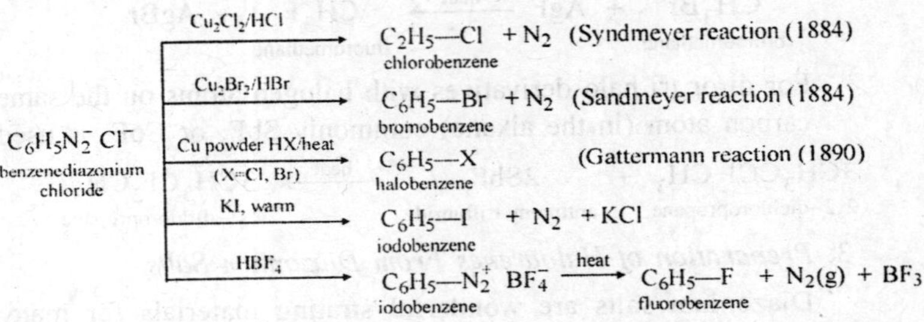

PHYSICAL PROPERTIES OF HALOALKANES AND HALOARENES

Some general physical properties of haloalkanes are described below.

(i) *Physical state and odour* : The lower members *e.g.*, bromomethane and chloroethane are gases. Iodomethane, and the majority of the higher members are sweet-smelling liquids. But still higher homologues are odourless solids.

Haloarenes are colourless liquids or crystalline solids.

(ii) *Colour* : Haloalkanes (alkyl halides) are colourless when pure, but bromides and particularly iodides develop colour when exposed to light.

(iii) *Boiling points* : The boiling points of halogen derivatives of hydrocarbons are influenced by the polarity of the C—X bond. Due to the greater polarity as well as higher molecular mass as compared to the parent hydrocarbon, the intermolecular forces of attraction are stronger in the halogen derivatives relative to that in hydrocarbons. As a result.

The boiling points of halogen derivatives are considerably higher than those of the corresponding hydrocarbons.

(i) If high temperature is avoided, no carbon selection rearrangement occur in either thermal or photochemical chlorination.

(ii) Every possible monochloride is formed, and over-chlorination, *i.e.*, chlorination beyond mono-substitution, may be suppressed by controlling the ratio of chlorine to alkane (this rule also holds good for photochemical chlorination).

(iii) The order of ease of substitution is tertiary hydrogen > secondary > primary. At 3000°C, with reaction in the vapour phase, the relative rates of substitution of primary secondary and tertiary hydrogen atoms are 1.00 : 3.25 : 4.43.

(iv) As the temperature rises above 300°C the relative rates of substitution tend to become equal, *i.e.*, 1 : 1 : 1. Increased pressure causes in the relative rate of primary substitution.

(v) According to Tender et at. (1960), the effect of the halogen atom already present in the molecule a to retard substitution at a β-carbon atom, and also affects substitution at an α-carbon atom.

It is of interest here to consider the method of calculation of rates of substitution. Let r_r r_s and r_t be the *relative* rates of substitution of a primary, secondary and tertiary hydrogen atom, respectively. Hass found that the thermal chlorination of propane at 300°C gave 48 per cent of the 1-chloro- and 52 percent of the 2-chloro-isomer. Since there are six H_p atoms, and statistically each one has an equal chance of being replaced, their total rate of replacement is 6r. Similarly for two H_s atoms their total rate of replacement is 2rs. Since the amount of each isomer formed depends on its rate of formation it therefore follows that

$$\frac{2r_s}{6r_p} = \frac{52}{48} \quad \therefore \quad \frac{r_q}{r_p} = \frac{3.25}{1}$$

Hass also found that chlorination of isobutane gave 67 per cent of the 1-chloro and 33 per cent of the 2-chloro-isomer. Since isobutane contains $9H_p$, atoms and H_t atom,

$$\frac{r_t}{9r_p} = \frac{33}{67} \quad \therefore \quad \frac{r_t}{r_p} = \frac{4.43}{1}$$

Thus the relative rates of substitution of H_p, H_s and H_t are 1 : 3. 25 : 4.43 (these are the figures given in (iii) above).

With these values it is possible to calculate the proportions of the various monochloro isomers for any alkane, *e.g.*, isopentane.

Structure	%	1	2	3	4
$(CH_3)_2\overset{2}{C}H\overset{3}{C}H_2\overset{4}{C}H_3$	calc.	30	22	33	14
	obs.	33.5	22	38	16.5

There are $9H_p$, $2H_s$ and $1H_t$. Thus the total rate is.

$$9 \times 1 + 2 \times 3.25 + 1 \times 4.43 = 19.93$$

$$\therefore \quad r_p \times 100 = \frac{9 \times 100}{19.93} = 45\%, \; r_s \times 100 = \frac{6.5 \times 100}{19.93} = 33\%,$$

$$r_t \times 100 = \frac{4.43 \times 100}{19.93} = 15\%\cdot$$

Since there are two CH_3 groups labelled 1 and one CH_3 group labelled 4, the rate of formation of the 1-chloro- derivative will be twice that of the 4-chloro. Thus 30 per cent will be the 1- and 15 per cent the 4- chloro isomer. These calculated values compare favourably with the values observed by Hass.

Bromination takes place with greater difficulty than chlorination, and there is less tendency for poly substitution. Furthermore, Roberts *et al*. (1952) have shown that bromination is more relective, bromine has much greater tendency than chlorine to replace tertiary hydrogen in preference to secondary or primary. It is the study of reactions such as these where the more reactive reagent shows less selectivity that has given rise to the empirical *reactivity selectivity principle.*

Pease *et al*. (1931) investigated the thermal chlorination of methane and found that the reaction was retarded by the presence of oxygen. This is strong evidence for a free-radical chain reaction and the authors proposed the following mechanism (cf. photohalogenation).

$$Cl_2 \xrightarrow{\text{heat}} Cl^\cdot + Cl^\cdot$$

$$Cl. + CH_4 \longrightarrow HCl + CH_3, etc..$$

Estimations have shown that in thermal chlorination of some alkanes, one chlorine atom may propagate a chain reaction involving about 10^7 hydrocarbon molecules.

Pease *et al*. also had proposed an alternative mechanism.

$$Cl. + CH_4 \longrightarrow CH_3Cl + H.$$

$$H^\cdot + Cl \longrightarrow HCl + Cl, \text{ etc.}$$

When we consider the enthalpy changes in these two chain reactions, we find (using bond energies in .

(i) (a) Cl· + Me—H (+ 414.2) ⟶ H— Cl (– 426.8) + Me·

$\Delta H^{\theta} = -12.6$ kJ

(b) Me. + Cl—Cl (242.7) ⟶ Me—Cl (– 326.4) + Cl.

$\Delta H^{\theta} = -83.7$k J

(ii) (a) Cl· + Me—H (+ 414.2) ⟶ Me—Cl (– 326.4) + H·

$\Delta H^{\theta} = +87.8$ kJ

(b) H· + Cl—Cl (+ 242.7) ⟶ H—Cl (– 426.8) + Cl·

$\Delta H^{\theta} = -184.1$ kJ

The overall enthalpy change for both mechanism is – 96.3 kJ. This of course, is to be expected since the overall equation for both mechanism (obtained by adding the steps of the reaction) is the same (Hess's law), viz.,

Cl—Cl (1 + 242.7) + Me—H (+ 414.2) ⟶Me— Cl (– 326.4)
+ H—Cl (– 426.8)

$\Delta H^{\oplus} = -96.3$ kJ.

Now let us consider the entropy change for the overall reaction values

(Jk^{-1} mol^{-1}) are. Cl_2 223.0, CH_4, 186.2, CH_3Cl. 234.3. HCl, 186.6.

$\therefore \quad \Delta S^{\oplus} = (234.3 + 186.6) - (223.0 + 186.2) = 11.7$ kJ

Hence there is a very small increase in entropy.

$\Delta G^{\oplus} = \Delta H^{\oplus} - T\Delta S^{\oplus} = -96.3 - (298 \times 11.7/1000)$

$= -96.3 - 3.49 = -99.79$ kJ

$\therefore$ RT In 99.79. log K = 99.79/2.3 RT

$= 99.79/5.7 = 17.4$ K = – 10^{17} at 25°C.

Thus the overall reaction (which is independent of mechanism) is thermodynamically highly favourable.

In the foregoing account, $\Delta H^{\oplus}$ has been *estimated* from bond energies. This usually gives a reason ably good estimate. Entropy changes have been evaluated from a knowledge of absolute entropy values of the reactants and products. Unfortunately, may of these values are not available. Since their estimation is very difficult, one can assume that the entropy changes are small at 25°C and so may be neglected in comparison with $\Delta H^{\oplus}$ values if these are reasonably high (say, 40 kJ, or more). In this ease, then, $\Delta H^{\oplus}$ may be taken as a measure of $\Delta G^{\oplus}$

Although the above reaction is thermodynamically highly favourable, it is immeasurably slow in the dark at room temperature. As we have seen, rates depend on activation energy values. Since the light) must be supplied from outside the system because the value required, 242.kJ, is far too high in be reached through thermal collision at room temperature. In mechanism (i) both (a) and (b) are exothermic, and so is thermodyanamically highly favourable. Now in an endothermic reaction, *E* must be at equal to ΔH^{θ}. Thus a minium value of *E* for (ii) is 87.86.k J.

This step is a therefore energetically highly unfavorable, and consequently, for a chain mechanism, (i) is involved rather than (ii). There is, however, one further point that should also be considered. It can be shown, from thermodynamic considerations, that the value of the equilibrium constant, K, decreases will rise in temperature for exothermic reactions and increases for endothermic reactions. This, of course follows from Lc Chatelier's principle. Thus, exothermic reactions become thermodynamically less favourable and endothermic reactions more favourable as the temperature rise, but in both cases the equilibrium position is reached much more quickly. Hence endothermic reactions can give reasonable yields at a fast rate when the temperature is raised sufficiently,. From this it follows that at some particular temperature, mechanism (ii) could compete with mechanism (i).

It is also of interest to consider the other halogens, fluorine, bromine and iodine.

	F	Cl	Br	I
$X_2 \longrightarrow 2X.$	+ 150.6	+212.7	+188.3	+150.6
(ia) $X^{\cdot}$ + MeH $\longrightarrow$ HX+Me	– 146.4	–12.55	+ 50.21	+117.2
(ib) $Me^{\cdot}$ + $X_2 \longrightarrow$ MeX+X.	– 334.7	–83.68	–96.23	–62.76
ΔH^{θ} (overall)	– 481.1	–96.23	–46.02	+54.44

The rate of each step will depend on the relevant activation energy. E for fluorination (ia) is apparently less than that for chlorination, and because the reaction is so strongly exothermic, fluorination is often explosive. E for bromination (ia) is about 74.48. kJ. The weaker the bond (that is broken), the lower is the necessary activation energy, and so the rate of substitution will be $\tau > s > p$. For bromination, the activation energy is reached for t-hydrogen, and so this is replaced almost exclusively, A might also be noted that order of replacement, τ > s p,. is what we might expect from the consideration of the stabilities of the free radicals produced in step (ia). Finally, as the temperature rise above 300°C (in thermal chlorination), the chlorine and alkane molecules acquire so much energy that reaction occurs at every collision respective of the type of C—H bond. Under these conditions, the replacement rates will be the same.

2. *Hunsdiecker or Brodine-Hunsdiecker reaction* : Hunsdiecker *et al.* (1935) found that silver and of the carboxylic acids in carbon tetrachloride solution are decomposed by chlorine of bromine to form the alkyl halide, *e.g.*,
3. Direct chlorination of alkanes may be effected by means of sulphuryl chloride. Sulphuryl chloride in the absence of light and catalysts does not react with alkanes even at their boiling points, but in the presence of light and a trace of an organic peroxide the reaction is fast. The mechanism is probably.

$$(R^1CO_2)_2 \longrightarrow 2R^1CO_2\cdot \longrightarrow 2R^1 + 2CO_2$$

$$R^{1\cdot} + SOlCl_2 \longrightarrow R^1Cl + SO_2Cl \longrightarrow SO_2 + Cl.$$

$$Cl:+R^2H \longrightarrow HCl + R^2. \xrightarrow{SO_2Cl_2} R^2Cl + SO_2Cl$$

In practice, chlorination with sulphuryl chloride is best carried out in the presence of αα–azobisisobutyronitrile (this readily produces free radicals).

$$\underset{\displaystyle CN}{\underset{|}{Me_2C}}—N{=}N—\underset{\displaystyle CN}{\underset{|}{CMe_2}} \longrightarrow N_2—2Me_2C(CN)\cdot$$

$$Me_2C(CN)\cdot + SO_2Cl_2 \longrightarrow Me_2C(CN).$$

Sulphuryl chloride is less selective than chlorine, and Russell (1958) has shown this is due to the nature of the solvent (when one is used).

CHEMICAL REACTIONS OF HALOGEN DERIVATIVES OF HYDROCARBONS

Reactivity of Haloalkanes

Haloalkanes are very reactive compounds. That is why these are used as the starting materials the preparation of a large variety of organic compounds. The high reactivity of haloalkanes is due to the polarity in the C—C bond.

The polarity of the C—X bond is due to the difference in the electronegativities of C and X atoms. The halogen atom (X) being more electronegative than X atom, pulls the shared pair of electrons towards itself. This causes a partial positive charge on the C atom, and a partial negative charge on X atom. This charge-separation is called *polarisation.* The polarity in the C—X bond is shown in Fig. 3.6. The extend of polarisation depends upon the nature of the halogen atom X.

$$—\overset{|}{\underset{|}{C}}{}^{+\delta} \longrightarrow X^{-\delta}$$

Fig. 3.6 : The polarity in the C—X bond is haloalkanes. This polarisation of bond is due to the difference between the electronegativities of C and X (halogen0 atoms.

The reactivity of haloalkanes depends upon the following factors.

1. *Nature of the halogen atom :* Reactivity of a haloalkane depends upon the bond energy of the C—X bond. Lower is the bond energy, greater is the reactivity of the haloalkane. Since, the bond energy of the C—X bond follows the order.

	C – Cl	>	C – Br	>	C – I
Bond energy, kJ, mol^{-1}	326.4		284.5		213.5

Hence, the reactivity of haloalkanes follows the order,

Iodoalkane > Bromoalkane > Cloroalkane

This appears logical because greater is the size of the halogen atom, greater is the ease with which it can be lost by the haloalkane.

2. *Nature of the alkyl group viz, primary, secondary and tertiary:* Haloalkanes are polar in nature — the alkyl group at the positive and the halogen at the negative end of the dipole. The alkyl groups are electron-releasing. So, the presence of an alkyl group increases the polarisation of the C — X bond.

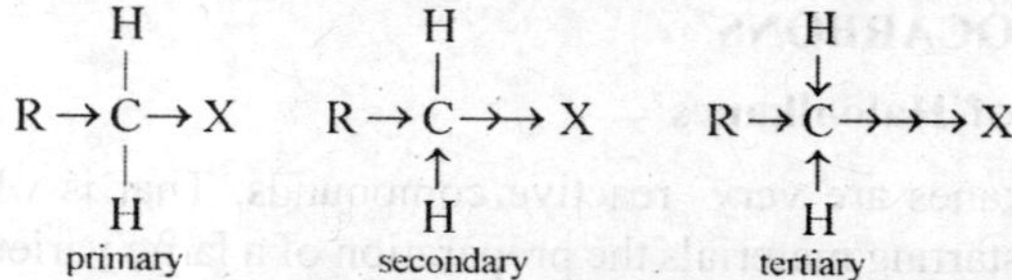

Large the number of alkyl groups attached to a carbon, greater is the repulsion of the electron-pair shared between C and X atoms towards the halogen atom. As a result, a tertiary haloalkane is more reactive than secondary haloalkane, and secondary haloalkane is more reactive than primary haloalkane. Thus the order of reactivity of haloalkanes is,

Tertiary haloalkane > Secondary haloalkane > Primary haloalkane

REACTIVITY OF HALOARENES RELATIVE TO HALOALKANES

Haloalkanes and haloarenes are polar compounds. The polarisation of C – X bond takes place towards the halogen atom due to its higher electronegativity, as shown below.

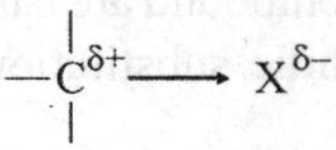

$$-\overset{|}{\underset{|}{C}}{}^{\delta+} \longrightarrow X^{\delta-}$$

As a result, the halogen atom becomes reactive. However, *haloarenes are less reactive than haloalkanes because of their stabilsation due to resonance.* The resonance in chlorobenzene is shown below in Fig. 17.2

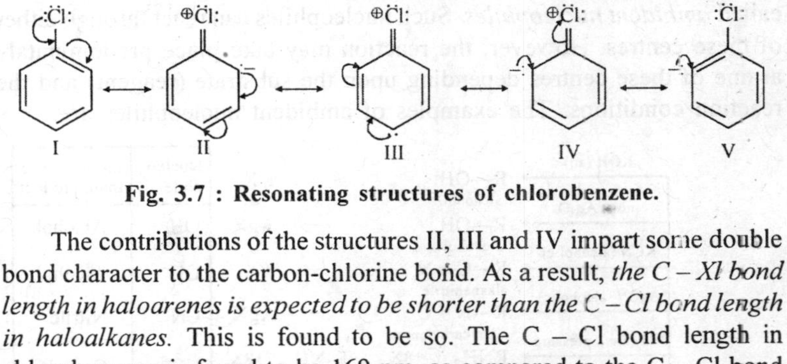

Fig. 3.7 : Resonating structures of chlorobenzene.

The contributions of the structures II, III and IV impart some double bond character to the carbon-chlorine bond. As a result, *the C – Xl bond length in haloarenes is expected to be shorter than the C – Cl bond length in haloalkanes.* This is found to be so. The C – Cl bond length in chlorobenzene is found to be 169 pm, as compared to the C – Cl bond length of 180 pm in chloroethane. This shortening of C – Cl bond length in haloarenes imparts stability to haloarenes chloroethane. This shortening of C – Cl bond length in haloarenes imparts stability to haloarenes relative to haloalkane. As a result, therefore, haloarenes are less reactive than haloalkanes.

CHEMICAL REACTIONS OF HALOGEN DERIVATIVES OF HYDROCARBONS

The characteristic reactions of halogen derivatives of hydrocarbons are described below.

Nucleophilic Substitution Reactions

Nucleus-loving, or electron-donating species are called *nucleophiles*. Nucleophiles may be natural molecules (Nu^-) with a lone-pair of negatively charged species ($\dot{N}u$). For example,

Neutral nucleophile: $H_2\ddot{O}:$ $:NH_3$, $R\ddot{O}H$, $R\ddot{N}H_2$

Negative nucleophile: $:\ddot{O}H^-$ X^- (X = F, Cl, Br), $CN:^-$,

Carbanions $-\overset{|}{\underset{|}{C}}:^-$

Nucleophiles attack the substrate at a location where the electron density is low. The reactions in which a strong nucleophile displaces a weaker nucleophile from a compound are called *nucleophilic substitution (SN) reactions. A* nucleophilic substitution reaction is represented as,

$$R-X \quad \underset{\text{nucleophile moves in}}{\underset{\text{stronger}}{Y^{\ominus}}} \longrightarrow R-Y + \underset{\text{nucleophile goes out}}{\underset{\text{wea ker}}{X^{\ominus}}}$$

The nucleophiles that have two nucleophilic centre of attack are called *ambident nucleophiles.* Such nucleophiles can react through either of these centres. However, the reaction may take place predominataly at one of these centres depending upon the substrate (reagent) and the reaction conditions. The examples of ambident nucleophiles are.

R—X (haloalkane):

- $\xrightarrow{\text{KOH (aq)}}$ R—OH (alchohol) + KX
- $\xrightarrow{\text{moist } Ag_2O}$ R—OH (alchohol) + AgX
- $\xrightarrow{\text{KCN (alcoholic)}}$ R—C ≡ N (alkanenitrile) + KX
- $\xrightarrow{\text{AgCN (alcoholic)}}$ R—N $\overset{\rightarrow}{=}$ C (alkyl carbylamine) + AgX
- $\xrightarrow[\text{-HX}]{NH_3\text{(alc.), pressure}}$ RNH_2 (pri.amine) $\underset{+HX}{\overset{+RX}{\longrightarrow}}$ R_2NH (sec.amine) $\underset{+HX}{\overset{+RX}{\longrightarrow}}$ R_3N (tert.amine) $\xrightarrow{+RX}$ $R_4N^+X^-$ (quaternary ammonium salt)
- $\xrightarrow[\text{(sodium alkoxide)}]{\text{R'ONa}}$ R—O—R′ (ether) + NaX
- $\xrightarrow{KNO_2}$ R—O—N = O (alkyl nitrite) + KX
- $\xrightarrow{AgNO_3}$ R—NO_2 (nitroalkane) + AgX
- $\xrightarrow{\text{NaSR'}}$ R—S—R′ (thioether) + NaX

Nucleo-phile	family of the main product
OH^-	Alcohol
OH^-	Alcohol
CN^-	Nitrile
NC^-	Carbyla-mine
$\dot{N}H_3$	amine
OR^-	Ether
O—N=O	Nitrite
NO_2^-	Nitro
SR^-	Thioether

- Cyanide ion ($^{-}$C ^{oo} N:) can attack through C or N
- Nitrite ion ($^{-}$O – $\dot{N}$ = O) can attack either through O or N.

Nucleophilic substitution reactions are used to prepare a variety of organic compounds.

Nucleophilic substitution reactions of haloalkanes : Nucleophilic substitution reactions given by haloalkanes are described below.

Nucleophilic substitution reactions of haloarenes : In haloarenes, the halogen atom is attached to the benzene ring, and it cannot be easily displaced by a nucleophilic agent (*e.g.*, OH^-, NH_3, CN^- etc.–.).

Thus, haloarenes are less reactive reactions of chlorobenzene are summarized below.

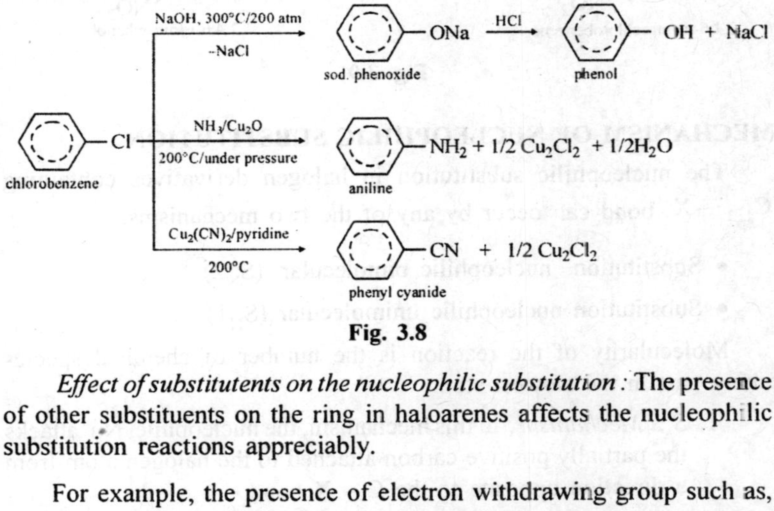

chlorobenzene —Cl —(NaOH, 300°C/200 atm, –NaCl)→ sod. phenoxide —ONa —(HCl)→ phenol —OH + NaCl

chlorobenzene —Cl —(NH_3/Cu_2O, 200°C/under pressure)→ aniline —NH_2 + 1/2 Cu_2Cl_2 + 1/2H_2O

chlorobenzene —Cl —($Cu_2(CN)_2$/pyridine, 200°C)→ phenyl cyanide —CN + 1/2 Cu_2Cl_2

Fig. 3.8

Effect of substitutents on the nucleophilic substitution : The presence of other substituents on the ring in haloarenes affects the nucleophilic substitution reactions appreciably.

For example, the presence of electron withdrawing group such as, – NO_2, – CN, – COOH etc., at *ortho–* and *para–* positions to the halogenation greatly activates the halogen atom for nucleophilic substitution reactions.

The effect of the presence of – NO_2 group on the reaction conditions for nucleophilic substitution reactions is illustrated below.

The presence of group, such as – NO_2, – CN, – COOH etc. on the ring makes the nucleophilic substitution easier.

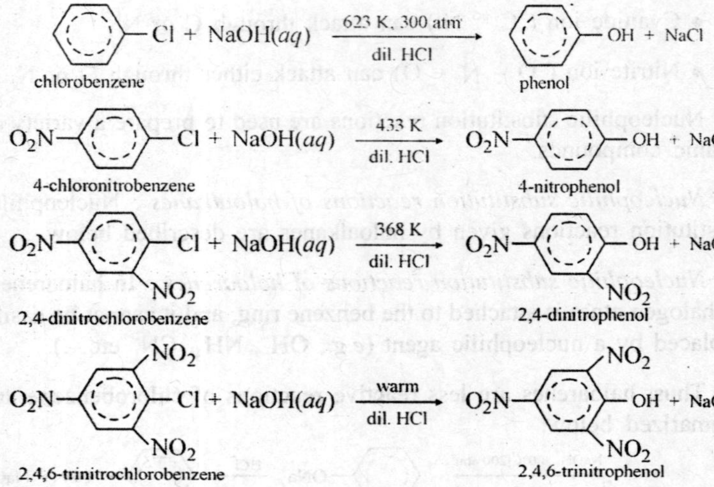

chlorobenzene + $NaOH(aq)$ $\xrightarrow[\text{dil. HCl}]{623\text{ K, }300\text{ atm}}$ phenol (–OH) + NaCl

4-chloronitrobenzene (O_2N–C_6H_4–Cl) + $NaOH(aq)$ $\xrightarrow[\text{dil. HCl}]{433\text{ K}}$ 4-nitrophenol (O_2N–C_6H_4–OH) + NaCl

2,4-dinitrochlorobenzene + $NaOH(aq)$ $\xrightarrow[\text{dil. HCl}]{368\text{ K}}$ 2,4-dinitrophenol + NaCl

2,4,6-trinitrochlorobenzene + $NaOH(aq)$ $\xrightarrow[\text{dil. HCl}]{\text{warm}}$ 2,4,6-trinitrophenol + NaCl

Fig. 3.9

MECHANISM OF NUCLEOPHILIC SUBSTITUTION

The nucleophilic substitution in halogen derivatives containing $C_{sp^3} - X$ bond can occur by any of the two mechanisms.

- Substitution nucleophilic bimolecular, (S_N2)
- Substitution nucleophilic unimolecular (S_N1).

Molecularity of the reaction is the number of chemical species taking part in the slowest stop.

1. *S_N2 Mechanism :* In this mechanism, the nucleophile, Nu^- attacks the partially positive carbon attached to the halogen atom from a direction opposite to the C – X bond.

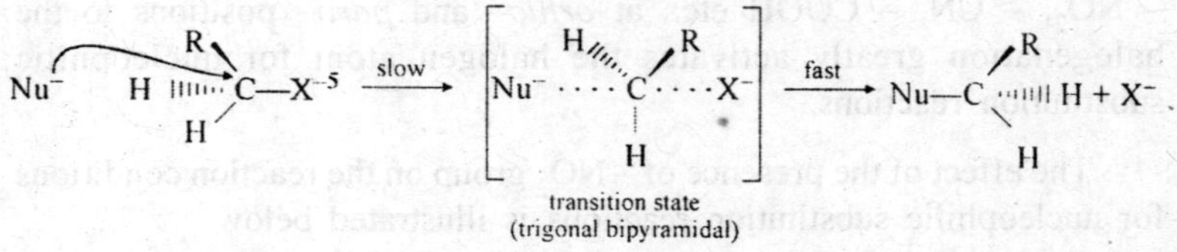

$$Nu^- \quad H\text{-}C(R)(H)\text{—}X^{-5} \xrightarrow{\text{slow}} \left[Nu^- \cdots C(H)(R)(H) \cdots X^- \right] \xrightarrow{\text{fast}} Nu\text{—}C(R)(H)H + X^-$$

transition state (trigonal bipyramidal)

Fig. 3.10

2. *S_N1 Mechanism :* In this mechanism, the reaction proceeds through the heterolytic fission of the C – X bond leading to the

formation of a carbocation, and a halide ion. The nucleophile then attacks the carbocation to form the substituted product. In this mechanisms, the first step is slow and the second step is fast.

The S_N1 mechanism is facilitated by polar solvents such as water, alcohol or aqueous–organic solvents.

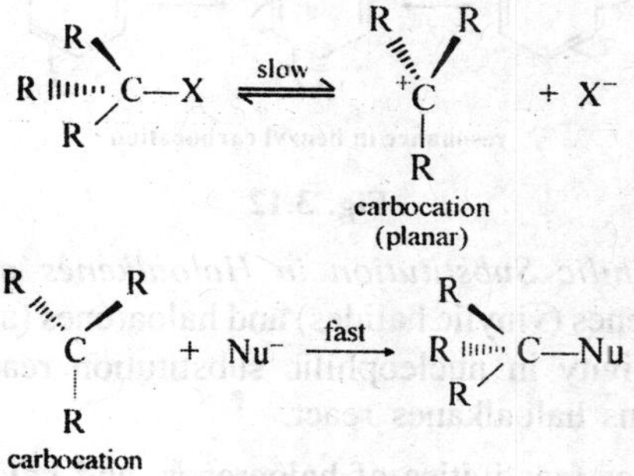

$$R_3C{-}X \xrightleftharpoons{\text{slow}} R_3\overset{+}{C} + X^-$$

carbocation (planar)

$$R_3\overset{+}{C} + Nu^- \xrightarrow{\text{fast}} R_3C{-}Nu$$

carbocation

Fig. 3.11

The order S_N1 reactivity is

$$CH_3{-}X < R{-}CH_2{-}X < R{-}\underset{R}{CH}{-}X < R{-}\underset{R}{\overset{R}{C}}{-}X$$

or primary < secondary < tertiary

The order S_N2 reactivity is

$$R{-}\underset{R}{\overset{R}{C}}{-}X < R{-}\underset{R}{CH}{-}X < RCH_2{-}X < CH_3{-}X$$

or tertiary < secondary < secondary

3. *S_N1 Mechanism in Primary Allylic and Primary Benzylic Halides:* In the S_N1 mechanism, the carbocation formed in the first step immediately reacts with the nucleophile to complete the reaction. More easily the carbocation is formed, easier is the overall reaction to occur. Thus, the tendency of the primary haloalkanes towards SN1 mechanism is quite low. However, the primary allylic and benzylic halides show greater reactivity than other primary haloalkanes in S_N1 reactions due to greater stabilization of the allylic and benzylic carbocations by resonance as shown below.

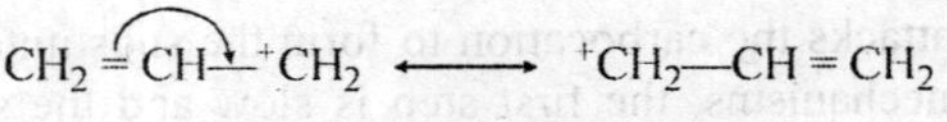

$CH_2 = CH - {}^+CH_2 \longleftrightarrow {}^+CH_2 - CH = CH_2$

resonance in allyl carbocation

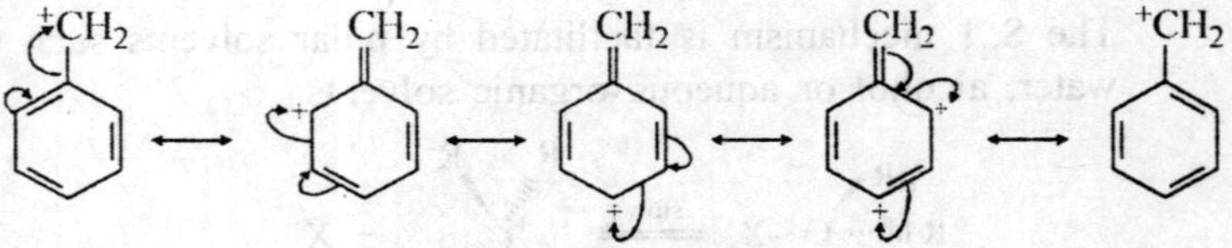

resonance in benzyl carbocation

Fig. 3.12

4. *Nucleophilic Substitution in Haloalkenes and Haloarenes:* Haloalkenes (vinylic halides) and haloarenes (aryl halides) show no reactivity in nucleophilic substitution reactions under the conditions haloalkanes react.

   The lower reactivities of haloarenes, and haloalkenes towards nucleophilic substitution is due to the delocalization of electrons by resonance imparting a partial double bond character in the carbon–halogen (C – X) bond.

   This shortens the C – X bond as compared to C – X single bond. As a result C – X bond in haloarenes and haloethenes becomes less reactive as compared to C_{sp^3}—X bond in haloalkanes.

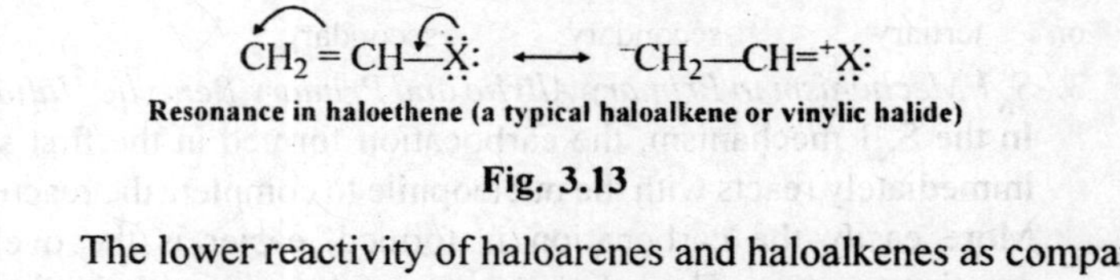

Resonance in halobenzene (a typical haloarene)

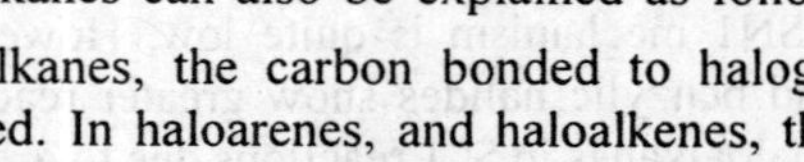

$CH_2 = CH - \ddot{X}: \longleftrightarrow {}^-CH_2 - CH = {}^+X:$

Resonance in haloethene (a typical haloalkene or vinylic halide)

Fig. 3.13

The lower reactivity of haloarenes and haloalkenes as compared to haloalkanes can also be explained as follows.

In haloalkanes, the carbon bonded to halogen atom is sp3 hybridized. In haloarenes, and haloalkenes, the carbon linked to halogen is sp2 hybridized. So, the carbon–halogen bond in haloarenes and haloalkenes is shorter and stronger due to some double bond character in it. As a result, *haloarenes and haloalkenes are more stable, and less reactive than haloalkanes.*

5. *Effect of Substituent on the Nucleophilic Substitution in Halobenzenes :* Haloarenes are not reactive in nucleophilic substitution reactions. However, haloarenes (in particular halobenzenes) become reactive to nucleophilic substitution when electron-withdrawing groups such as, – NO_2, – CN etc., are present at the *ortho* and *para* positions. The reactivity increases progressively with the number of such groups.

The increase in the reactivity gets reflected in the progressively milder conditions required for the nucleophilic reaction to take place. For example, the effect of – NO_2 group on the hydrolysis of chlorobenzene can be seen in the reaction conditions.

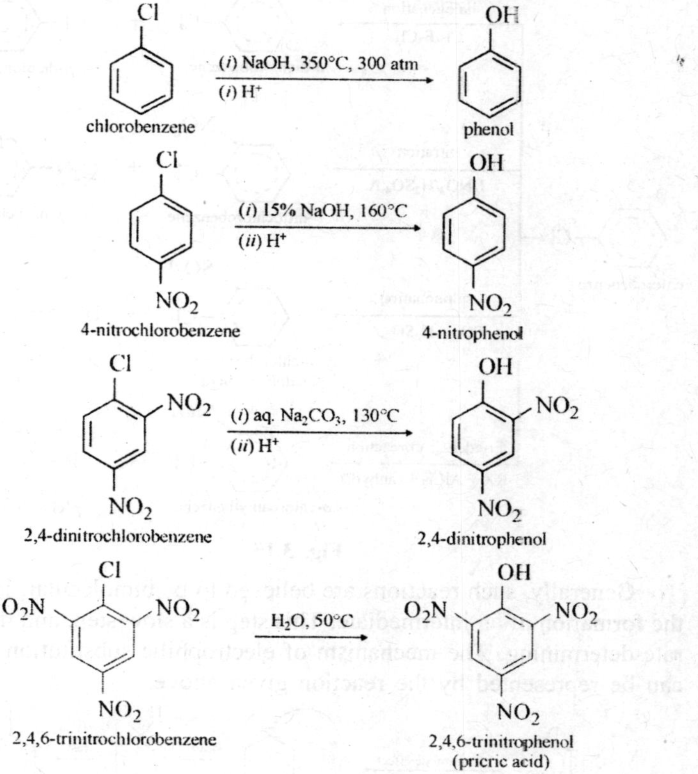

chlorobenzene $\xrightarrow[(i)\ H^+]{(i)\ NaOH,\ 350°C,\ 300\ atm}$ phenol

4-nitrochlorobenzene $\xrightarrow[(ii)\ H^+]{(i)\ 15\%\ NaOH,\ 160°C}$ 4-nitrophenol

2,4-dinitrochlorobenzene $\xrightarrow[(ii)\ H^+]{(i)\ aq.\ Na_2CO_3,\ 130°C}$ 2,4-dinitrophenol

2,4,6-trinitrochlorobenzene $\xrightarrow{H_2O,\ 50°C}$ 2,4,6-trinitrophenol (pricric acid)

Fig. 3.14

ELECTROPHILIC SUBSTITUTION REACTIONS IN HALOARENES

Due to the presence of electron-rich benzene (or aromatic) ring, the haloarenes undergo electrophilic substitution reactions.

The halogen atom directs the incoming substituent to the *ortho,* and *para* position relative to the halogen atom. Due to – I effect, halogen atom deactivates the benzene ring. As a result, the electrophilic substitution reactions of haloarenes occur slowly and under drastic conditions as compared to the benzene.

Any chemical species seeking electrons is called an *electrophile.* Electrophiles attack a centre of increased electron-density in any molecule, such as benzene ring.

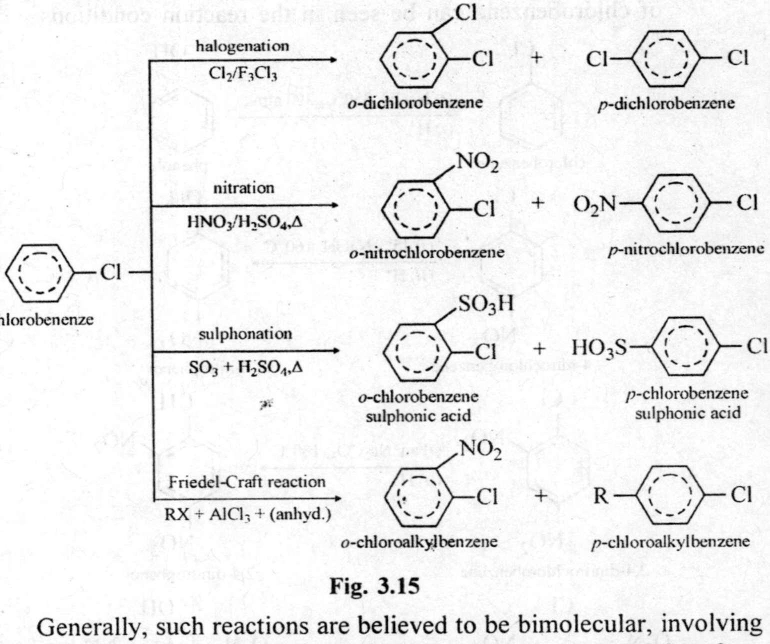

Cl
halogenation
Cl_2/F_3Cl_3
Cl
+
Cl
Cl
o-dichlorobenzene
p-dichlorobenzene
NO_2
nitration
$HNO_3/H_2SO_4,\Delta$
Cl
+
O_2N
Cl
o-nitrochlorobenzene
p-nitrochlorobenzene
Cl
chlorobenenze
SO_3H
sulphonation
$SO_3 + H_2SO_4,\Delta$
Cl
+
HO_3S
Cl
o-chlorobenzene sulphonic acid
p-chlorobenzene sulphonic acid
NO_2
Friedel-Craft reaction
$RX + AlCl_3$ + (anhyd.)
Cl
+
R
Cl
o-chloroalkylbenzene
p-chloroalkylbenzene

Fig. 3.15

Generally, such reactions are believed to be bimolecular, involving the formation of an intermediate. This step is a slow step, and therefore, rate-determining. The mechanism of electrophilic substitution reaction can be represented by the reaction given above.

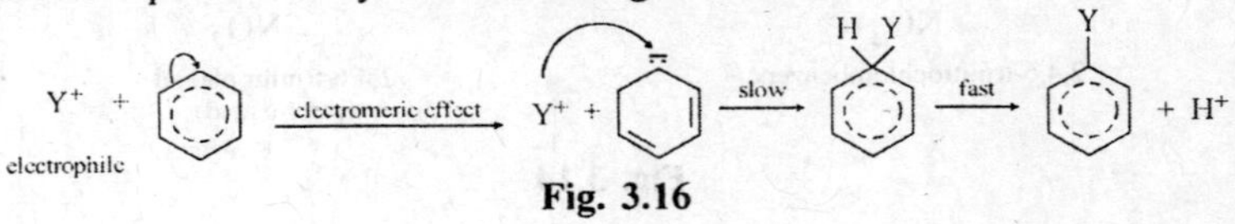

Y^+ +
electrophile
electromeric effect
Y^+ +
slow
H Y
fast
Y
+ H^+

Fig. 3.16

ELIMINATION REACTIONS

An elimination reaction involves the removal or one molecule of HX from the molecule of a haloalkane. The elimination reactions, therefore, are also called *dehydrohalogenation reactions.*

In elimination reactions the halogen atom of the haloalkane, and a hydrogen atom form the adjacent carbon atom (β-carbon) are involved. That is why, the elimination reactions of haloalkanes are called *β-elimination reactions.* The order of reactivity in elimination reactions is

Chloro < Bromo < Iodo

Primary < Secondary < Tertiary

The reactivity of elimination reaction also increases with increasing number of substituents at the β-carbon.

Haloalkane when boiled with alcoholic potassium hydroxide form alkenes by the elimination of one molecule of HX.

$$\underset{\text{haloalkane}}{-\overset{|}{\underset{|}{{}^{\beta}C}}-\overset{|}{\underset{|}{{}^{\alpha}C}}-}\ (\boxed{H-X}) + KOH\,(alc.) \xrightarrow{\text{boil}} \underset{\text{alkene}}{-\overset{|}{C}=\overset{|}{C}-} + KBr + H_2O$$

Example:

Bromomethane on boiling with alcoholic KOH, gives ethene.

$$H-\overset{H}{\underset{H}{\overset{|}{\underset{|}{C}}}}-\overset{H}{\underset{Br}{\overset{|}{\underset{|}{C}}}}-H + KOH\,(alc) \xrightarrow{\text{boil}} \underset{\text{ethene}}{H-\overset{H}{\overset{|}{C}}=\overset{H}{\overset{|}{C}}-H} - KBr + H_2O$$

(b) 1–bromopropane with KOH (alc), gives propene.

$$\underset{\text{1-bromopropane}}{H-\overset{H}{\underset{H}{\overset{|}{\underset{|}{C}}}}-\overset{H}{\underset{H}{\overset{|}{\underset{|}{C}}}}-\overset{H}{\underset{Br}{\overset{|}{\underset{|}{C}}}}-H} + KOH\,(alc) \xrightarrow{\text{boil}} \underset{\text{propene}}{H-\overset{H}{\overset{|}{C}}=\overset{H}{\overset{|}{C}}-\overset{H}{\underset{H}{\overset{|}{\underset{|}{C}}}}-H} + KBr + H_2O$$

If a haloalkane can eliminate a HX molecule in more than one way, then the formation of the most substituted alkene' predominates. This is called *Satzeff's rule.*

Saytzeff rule : In dehydrohalogenation reactions, the preferred product is that alkene which has greater number of alkyl groups attached to the doubly bonded carbon atoms. For example 2-bromobutane (*see* butyl bromide) can eliminate one molecule of HBr in the following two ways.

$$CH_3-{}^{\beta}CH_2-{}^{\alpha}\underset{\displaystyle Br}{\underset{|}{C}}H-{}^{\beta}CH_3 \xrightarrow{KOH\,(alc)} CH_3-CH{=}CH-CH_3 + H_2O + KBr$$

but-2-ene (80%)
(preferred mode of elimination)
(here, a disubsituted alkene is formed)

$$\xrightarrow{KOH\,(alc)} CH_3-CH_2{=}CH-CH_2 + KBr + H_2O$$

but-1-ene (20%)
(here, a monosubstituted alkene is formed)

In this reactions, but-2-ene is a more substituted alkene than but-1-ene. Therefore, but-2-ene is the major product.

REACTIONS WITH METALS

Haloalkanes react with active metals to form products which depend upon the nature of the metal used.

(a) *With sodium metal :* Haloalkanes when heated with metallic sodium in ether solution given alkanes. This reactions is known as *Wurtz reaction.* This reactions is used for obtaining symmetrical alkanes. This reaction, however, fails with tertiary haloalkanes and usually a mixture of alkanes is obtained.

$$\underset{\text{haloalkane}}{2RX} + 2Na \xrightarrow{\text{ether}} \underset{\text{alkane}}{R-R} + 2NaX$$

$$RX + 2Na + R'X \xrightarrow{\text{ether}} R-R + R-R' + R'-R' + 2NaX$$

Example:

$$\underset{\text{chloroethane}}{2C_2H_5Cl} + 2Na \xrightarrow{\text{ether}} \underset{\text{n-bu tan e}}{C_2H_5 - C_2H_5} + 2NaCl$$

Haloarene when heated with sodium metal in presence of dry ether, the halogen atom of haloarene is replaced by an aryl group. For example, chlorobenzene with sodium in ether gives biphenyl. This reaction is called *Fitting reaction.*

$$\underset{\text{chlorobenzene}}{C_6H_5-Cl} + 2Na + \underset{\text{chlorobenzene}}{Cl-C_6H_5} \xrightarrow{\text{dry ether}} \underset{\text{biphenyl}}{C_6H_5-C_6H_5} + 2NaCl$$

Fig. 3.17

When a haloarene is heated with a haloalkane and sodium in ether, the halogen atom of haloarene is replaced by the alkyl group. This reaction is called *Wurtz-Fitting reaction*. For example,

Wurtz-Fitting reaction

$$C_6H_5-Cl + 2Na + Cl-CH_3 \xrightarrow{\text{dry ether}} C_6H_5-CH_3 + 2NaCl$$

chlorobenzene (haloarene) — chloromethane (haloalkane) — toluene

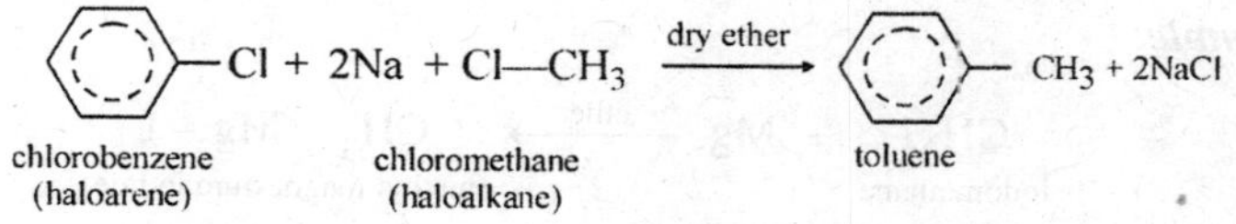

Fig. 3.18

(b) *When other active metals* : Haloalkanes also react with some other active metals such as, lithium, aluminium, zinc, and lead etc., in dry ether to form the corresponding organometallic compounds. For example,

$$CH_3-CH_2Br + 2Li \xrightarrow{\text{dry ether}} CH_3-CH_2Li + LiBr$$

bromoethane — ethyl lithium

$$CH_3-\underset{\displaystyle Br}{\underset{|}{CH}}-CH_3 + 2Li \xrightarrow{\text{dry ether}} CH_3-\underset{\displaystyle Li}{\underset{|}{CH}}-CH_3 + 2LiBr$$

2-bromopropane — 2-propyl lithium

$$2CH_3-CH_2Br + 2Zn \xrightarrow[\Delta,\text{ inert atm}]{\text{dry ether}} (CH_3-CH_2)_2Zn + ZnBr_2$$

bromoethane — diethyl zinc

$$4CH_3-CH_2Br + 4Na-Pb \longrightarrow (CH_3-CH_2)_4Pb + 4NaBr + 3Pb$$

teraethyl lead (TEL) (used as an antiknocking agent in gasoline)

Like haloalkanes, haloarenes also react with certain active metals. Bromoarenes and iodoarenes react with lithium in the presence of dry ether to give phenyl lithium.

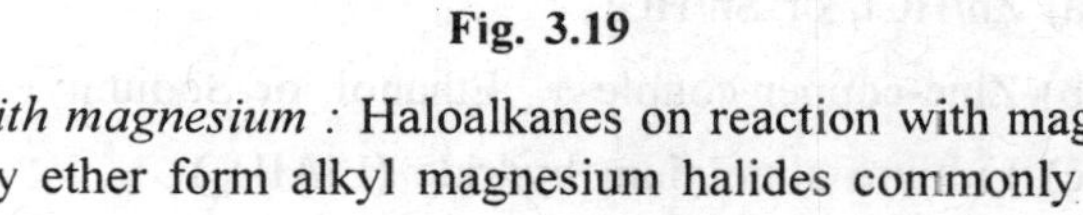

$$C_6H_5-Br + 2Li \xrightarrow{\text{dry ether}} C_6H_5-Li + LiBr$$

bromobenzene — phenyl lithium

Fig. 3.19

(c) *With magnesium* : Haloalkanes on reaction with magnesium in dry ether form alkyl magnesium halides commonly known as the *Grignard reagent,* (R – Mg – X). *Grignard reagent is an organometallic compound.*

$$RX + Mg \xrightarrow[\Delta]{\text{ether}} R - Mg - X$$

alkyl magnesium halide
(Grignard reagent)

Example:

$$\underset{\text{iodomethane}}{CH_3I} + Mg \xrightarrow[\Delta]{\text{ether}} \underset{\text{methyl magnesium iodide}}{CH_3 - Mg - I}$$

$$C_3H_7Br + Mg \xrightarrow[\Delta]{\text{ether}} C_3H_7MgBr$$

Bromo and iodoarenes react with dry magnesium power in the presence of dry ether to yield alymagnesium halides *(Grignard's reagent).*

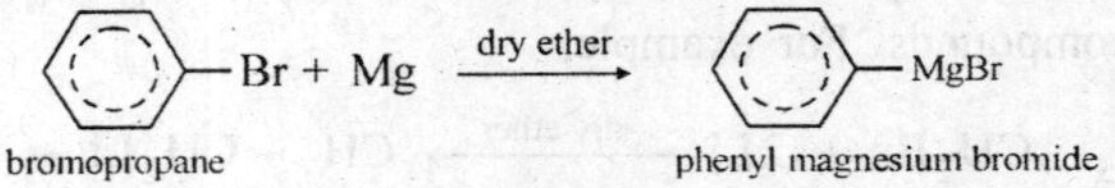

$$\underset{\text{bromopropane}}{C_6H_5-Br} + Mg \xrightarrow{\text{dry ether}} \underset{\text{phenyl magnesium bromide}}{C_6H_5-MgBr}$$

Fig. 3.20

Chloroarenes react with magnesium in the presence of tetrahydrofuran (THF) solvent to give phenylmagnesium chloride.

$$\underset{\text{bromopropane}}{C_6H_5-Cl} + Mg \xrightarrow{\text{THF}} \underset{\text{phenyl magnesium chloride}}{C_6H_5-MgCl}$$

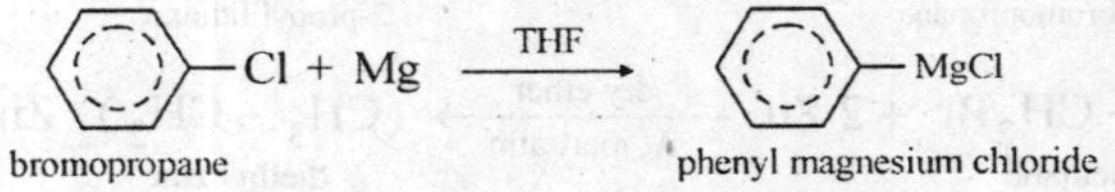

Fig. 3.21

Grignard reagents, (named after their discover Victor Grignard, a French chemist) are very useful preparative reagents. They react with any protic compound to form hydrocarbons.

$$R - Mg + X + ZH \ (Z = OH, RO, RNH, RC \equiv C) \longrightarrow RH + MgZX$$

REDUCTION REACTIONS

Haloalkanes on reduction give the corresponding alkanes. Reduction of haloalkanes can be carried out by using,

(a) Zn/HCl, or Sn/HCl

(b) Zinc-copper couple + Ethanol, or Sodium + Ethanol

(c) Lithium aluminium hydride ($LiAIH_4$)

$$\underset{\text{haloalkane}}{RX} + 2H \xrightarrow{Zn/HCl} \underset{\text{alkane}}{RH} + HX$$

Example:

$$\underset{\text{bromoethane}}{C_2H_5Br}\ \ 2H \xrightarrow{Zn-Cu/HCl} \underset{\text{ethane}}{C_2H_6} + HBr$$

Iodoalkanes may be reduced by heating it with HI and red phosphorus at 150°C.

$$\underset{\text{iodoethane}}{C_2H_5I} + HI + \underset{\text{red}}{P} \xrightarrow{150°C} \underset{\text{ethane}}{C_2H_6} + I_2$$

Haloarenes are reduced by nickel-aluminium alloy in alkali, or by sodium amalgam and aqueous alcohol in the presence of alkali, to produce the corresponding hydrocarbons. For example,

$$\underset{\text{chlorobenzene}}{C_6H_5-Cl} + 2H \xrightarrow{Ni\text{-}Al/alkali} \underset{\text{bensene}}{C_6H_6} + HCl$$

Fig. 3.22

REARRANGEMENT REACTIONS

Haloalkanes on heating in the presence of anhydrous aluminium chloride form an isomeric haloalkane. Such a rearrangement reaction is called *isomerisation*.

$$\underset{\text{1-chloropropane}}{CH_3-CH_2-CH_2Cl} \xrightarrow[3000°C]{AlCl_3\,(anhyd.)} \underset{\text{2-chloropropane}}{CH_3-\underset{\underset{Cl}{|}}{CH}-CH_3}$$

DISTINGUISHING TEST FOR HALOALKANES AND HALOARENES

Haloalkanes and haloarenes contain halogen atom in their molecules. To distinguish between the two *silver nitrate test* is conducted.

Test : Take about 1 mL of the given compound in a test tube. Add to it about 2 mL of aqueous KOH, and heat to boiling. Cool the solution and acidified with dil. HNO_3. To this, add a few drops of $AgNO_3$ solution, and observe for any white precipitate.

(a) Haloalkanes give white precipitate.

(b) Haloarenes and haloalkenes do not give white precipitate.

USES OF HALODERIVATIVES OF HYDROCARBONS

Haloalkanes are very reactive compounds. Therefore, haloalkanes are used as important starting materials for synthesizing a large number of organic compounds.

Haloarenes are used as the starting material for the manufacture of compounds, such as, phenol, aniline, D.D.T. etc.

GENERAL PROPERTIES OF THE ALKYL HALIDES

The lower members methyl chloride, methyl bromide and ethyl chloride are gases, methyl iodide and the majority of the higher members are sweet-smelling liquids. The order of values of the poiling points (and densities) of the alkyl halides is iodide > bromide > chloride > fluoride. In a group of isomeric alkyl halides, the order of the boiling points is primary > secondary > tertiary. Many of the alkyl halides burn with a green-edged flame. The chemical reactions of the alkyl halides are similar, but they are not equally reactive, the order of reactivity being iodide > bromide > chloride, the reactivity of alkyl fluorides depends on the nature of the fluoride. Alkyl iodides are sufficiently reactive to be decomposed by light, the iodide darkening due to the liberation of iodine.

$$2RI \longrightarrow R—R + I_2$$

Alkyl halides (and the polyhalides) are covalent compounds, insoluble in water, with which they cannot form a hydrogen bond, but soluble in organic solvents.

The alkyl halides are classified as primary, secondary and tertiary, according as the halogen atom is present in the respective groups — CH_2X =CHX and ≡CX, *i.e.*, according as the given atom is joined to a primary, secondary or tertiary carbon atom.

The absorption region of alkyl halides (C—X stretch) depends on the nature of the halogen atom, C—F, 1000 cm^{-1} (s), C—Cl, 750 –700 cm^{-1} (s), C—Br, 600–500 cm^{-1} (s), C—I, –500 cm^{-1} (s)

NMR spectroscopy of alkyl halides has been previously discussed. Here we shall mention the problem of fluorine. Apart from being the most electronegative atom of the halogen group, fluorine has nuclear spin, with I = + 1/2 and –1/2 (as for hydrogen). It can there for spin-spin couple with protons and so give rise to splitting, *e.g.*, CH_3F gives a doublet for the methyl group (τ 5.70), whereas CH_3Cl gives singlet (τ 6.98) and CH_3Br also a singlet (τ 7.35). The signal for fluorine is not observed under the conditions used for proton resonance, but with the conditions fulfilled, the fluorine spectrum will be a quartet.

Mass spectra. Loss of one electron from a lone pair on the halogen atom occurs most readily to form the molecular ion, and this then undergoes fragmentation in a way which depends on the nature of the halogen atom and on the nature of the alkyl group. Since Cl and Br exist as isotopes, the molecular ions appear as doublets of mass M and M + 2. However, because the abundance ratios are quite different, alkyl chlorides and bromides may be readily distinguished. Some of the fragmentation patterns are:

$$RCH_2CH_2^+ + X\cdot \xleftarrow{\text{Br. I}} [RCH_2CH_2\text{—}X]^+ \xrightarrow{\text{FeCl}} [RCH{=}CH_2]^+ + HX$$

$$[RCH_2CH_2\text{—}X]^+ \longrightarrow RCH_2^+ + CH_2{=}X^+$$

These are the predominant paths and the intensity of M^+ is in the order I > Cl > F, and decreases with increases in the size of the alkyl group. The $RCH_2CH_2^+$ and $[RCH{=}CH_3]^+$ fragments undergo further fragmentation typical of alkanes and alkenes, respectively. Alkyl halide (predominantly for Cl and Br) containing six or more carbon atoms in a straight chain also fragment to give the ion $C_4H_4X^+$.

$$R\text{—}CH_2\text{—}CH_2\text{—}CH_2\text{—}CH_2\text{—}X^+ \longrightarrow R\cdot + \text{cyclic } [CH_2CH_2CH_2CH_2X]^+$$

It has already been pointed out that alkyl groups have a + I effect. Before considering the reactions of the alkyl halides, it is important to explain why an alkyl group is electron-repelling. Several explanations have been offered the following is a highly favoured one. As we have seen methane and ethane are non-polar. Thus the methyl group has a zero inductive effect, and this is true for all alkyl groups since all alkanes, whether straight or branched-chain, have a into dipole moment. When, however, one]hydrogen atom in an alkane is replaced by some polar atom (or group), the alkyl group now exerts a polar effect which is produced by the presence of the polar atom. Thus, in an alkyl halide, the alkyl group possesses an inductive effect, but it is one which is produced mainly by the mechanism of *interaction polarisation.* It therefore follows that the alkyl group is more polarisable than a hydrogen atom, and since most of the groups attached to the alkyl group are electron-attracting groups, the alkyl groups thus usually becomes an electron-repelling group, *i.e.*, alkyl groups normally here a + 1 effect.

Optical isomerism : Optical activity is the name given to the phenomenon exhibited by compounds which, when placed in the path of a beam of plane-polarised. Light, are capable of getting the plane of polarisation to the left or right. Such compounds are said to be *optically* atoms, and are dextrorotatory if the substance rotates the plane of polarisation to the right and levorotatory if to the left. The symbols (+) and (–) are used to indicate the sign of rotation, *e.g.*, (+)-glucose (dextrorotatory). (–)-fructose (levorotatory).

Optical activity may be due to the nature of the crystalline structure of the compound to, but for the present brief discussion we shall consider optical activity that is due to nature of the *molecular* structure of the compound.

When a molecular structure is not supper impossible on its mirror image, that molecule is said to posses *molecular asymmetry,* and as result, exits in two forms, one being the (+)-from and the other the (–)-form, the rotation being numerically equal.

These two forms constitute a pair of enantiomorphs or enantion and a mixture of a pair of enantiomers in equimolecular amounts is called a racemic modification. This is optically inactive (by external compensation) and is represented as DL- or (±)-.

From the chemical point of view it is less important to know the signs of rotation than know *stereochemical relationships i.e.* the relationships between the configurations of various optically active compounds. These relationships are indicated by the symbols D and I, D (+)-glucose and D(–) -fructose are stereochemically related, *i.e.* their configurations are related (both are members of the D-series) even through their rotations happen to be optical sign.

A more detailed account of optical isomerism is given in Ch. 17, but it should be noted here that may molecule that contains an asymmetric carbon atom. *i.e.*, a carbon atom joined to for *different* groups, possesses molecular asymmetry.

Furthermore, substitution reactions occurring at an asymmetric carbon atom may be accompanied by *optical inversion (Walden investing i.e.*, a D-substrate, on undergoing reaction, produces the L-product, or vice versa.

GENERAL REACTIONS OF THE ALKAL HALIDES

1. Alkyl halides are hydrolysed to alcohols very slowly by water, but rapidly by silver oxyl suspended in boiling water, or by boiling aqueous alkalis (see also below).

$$RX + KOH \longrightarrow ROH + KX$$

This type of reaction is an example of nucleophilic substitution (S_N) since the attacking reagent is nucleophilic regent. It has been assumed that this type of heterolytic reaction in solution can take place by two different mechanisms, unimolecular or bimolecular.

Unimolecular mechanism : This is a *two-stage process*. The first stage consisting to slow heterolysis of the compound to form a carbonium ion, followed by rapid combination between the carbonium ion and the substituting nucleophilic reagent. Since the rate-determining step is the first one, and since in this step only one molecule is undergoing a covalency change, this type of mechanism is called *unimolecular* and is labelled S_N1 (Ingold *et al.* 1928. 1933). Thus, if the hydrolysis of an alkyl halide is an S_N1 reaction it may be written.

$$R\text{—}X \xrightleftharpoons{\text{slow}} X^{-}\text{—}R^{+} \xrightarrow[\text{fast}]{OH^{-}} ROH$$

Heavy metal ions, particularly silver ions, catalyse S_N1 reactions and according to Kevill *et al.* (1967), the mechanism involves a fast pre-equilibrium step (the silver ion has an empty orbital).

$$R\text{—}X + Ag^{+} \xrightleftharpoons{\text{heat}} [R\text{—}X\text{—}Ag]^{+} \xrightarrow{\text{heat}} R^{+} + AgX$$

Bimolecular mechanism : This is a *one-stage process,* two molecules simultaneously undergoing covalency change in the rate-determining step. This type of mechanism is called *bimolecular* and is labelled S_N2. Since the rate-determining step in this reaction is the formation of the transition state, the hydrolysis of an alkyl halide by an S_N2 reaction may be written.

$$HO\ R\text{—}X \xrightarrow{\text{slow}} \overset{\delta-}{HO} \cdots R \cdots \overset{\delta-}{X} \xrightarrow{\text{fast}} HO\text{—}R + X^{-}$$

A number of differences exist between S_N2 and S_N1 reactions, *e.g.*, (i). When both reactants are present in small and controllable concentrations. S_N2 reactions are second-order and S_N1 reactions are first-order. If one reactant is present in large excess *e.g.*, one of the reactants is the solvent, than an S_N2 mechanism will lead to first-order kinetics. (ii) The S_N2 mechanism always leads to inversion of the product, whereas with the S_N1 mechanism the configuration of the product depends on various factors, (iii) No rearrangement is possible for the S_N2 mechanism, but is possible (and usually occurs) for the S_N1, (iv) The rate constant of an S_N2 reaction with a given substrate depends on the nature of the nucleophilic and for a given nucleophile, on the nature of the leaving group in the substrate (see above).

Stereochemistry of Substitution Reactions

(i) *The S_N2 mechanism.*

$$HO^- \; R—X \longrightarrow \overset{\delta-}{HO} \cdots R \cdots \overset{\delta-}{X} \longrightarrow HO—R + X^-$$

In the transition state, the groups OH and X are collinear and on opposite sides of the attacked carbon atom. Furthermore, the line joining OH and X is prependicular to the plane containing the other three groups a, b, and d. In the original molecule *Cabd* X, the four groups are arranged tetrahedrally. Hence, to achieve a planar configuration of *Cabd* in the transition state, the carbon atom changes from tetrahedral to trigonal hybridisation, the remaining p_x orbital being used (by means of its two loves) to hold the groups OH and X by 'half-bonds'. When X is ejected, the carbon atom returns to its state of tetrahedral hybridisation.

The above reaction is a three-centre-reaction, proceeding by an end-one approach.

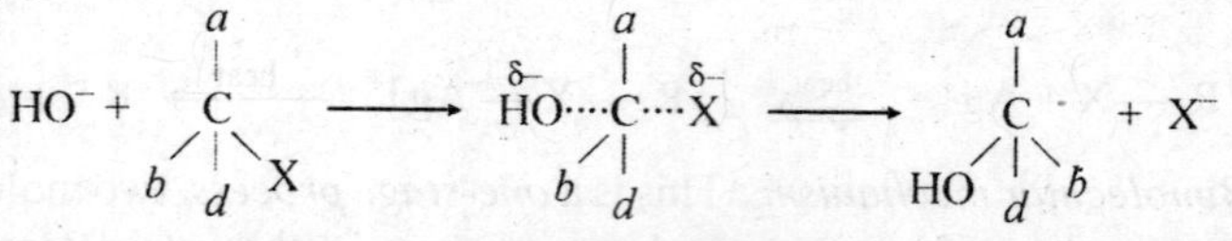

$$HO^- + C(a)(b)(d)X \longrightarrow \overset{\delta-}{HO} \cdots C(a)(b)(d) \cdots \overset{\delta-}{X} \longrightarrow HO\,C(a)(d)(b) + X^-$$

Hughes *et al.*, (1935) studied. (a) the interchange reaction of (+)-2iodo-octane with radio active iodine (as NaI*) in acetone solution, and (b) the racemisation of (+)-2-iodo-octane by ordinary NaI under the same conditions. These reactions were shown to occur by the S_N2 mechanism, and the result showed that every halide-halide* displacement was always accompanied by inversion. Thus this experiment leads to the *assumption that an S_N2 reaction always gives inversion,* this is fully supported by other experimental work. Hence, if it can be shown that the S_N2 mechanism is operating, then inversion must have occurred.

(ii) *The S_N1 mechanism*

$$R—X \underset{}{\overset{slow}{\rightleftharpoons}} X^- + R^+ \xrightarrow[fast]{OH} ROH$$

When the reaction proceeds by this mechanism, then inversion and retention (recemisation) will occur, the amount of each depending on various factors. The carbonium ion is flat (trigonal hybridisation), and hence attack by nucleophilic regents can take place equally well on either side, *i.e.*, equal amounts of (+)- and (–)-forms are produced, this is recemisation. Complete recemisation, however, can be expected only if the carbonium ion is sufficiently long-lived, this is favoured by low reactivity of the carbonium ion and low concentration of the nucleophilic reagent. On the other hand, during the actual ionisation, the retiring negative group will shield attack on that side, this encourages end on attack *on the other side*, there by leading to inversion. In general, the S_N1 mechanism is accompanied by inversion and racemisation, but in some cases there may be complete inversion.

It will be seen from the above discussion that if heterolysis of RX is the rate-determining step, than the mechanism, is S_N1. However, circumstances might arise where the *second* step is the rate-determining one. In this case, Hughes and Ingold (1954) have labelled the reaction S_N2 (C*) [substitution, nucleophilic, bimolecular, with a rapidly formed carbonium ion.

Another S_N reaction is the SN1 type (substitution, nucleophilic, internal), *e.g.*, Hughes. Ingold *et al* (1937) have shown that optically active α-phenylethanol (C6H5CHMe – = R) reacts with thionyl chloride to give a-phenylethyl chloride with almost complete retention of configuration. The mechanism proposed in S_N1 via a chlorosulphite.

$$R{-}O(H)\cdots Cl{-}S(Cl){=}O \xrightarrow{\text{fast}} HCl + R{-}O{-}S(Cl){=}O \xrightarrow{\text{fast}} RCl + SO_4$$

Now inversion occurs in either stage (both stages are four-centre type). The situation arises with *arylmethanols,* but in the presence of pyridine, the chloride now has the *inverted* configuration. This has been explained on the basis that the S_N2 mechanism is operating, but now involves a pyridine complex.

$$ROSOCl + C_5H_5N \xrightarrow{\text{fast}} ROSONC_5H_5 + \xrightarrow[S_N2]{\text{slow}}$$

$$\overset{\delta-}{Cl}...R...\overset{\delta-}{O}\,SONC_5H_5 \xrightarrow{\text{fast}} Cl-R + SO_2 + C_5H_5N.$$

Any factor that affects the energy of activation of a given type of reaction will affect the rate and/or the mechanism. The following discussion is largely qualitative, and because of this, one cannot be sure which are the predominant factors in deciding the value of the energy of activation (see also later).

Let us first consider the nature of R. There are two effects to be considered, the polar factor and the steric factor.

polar factor. Consider the series of ethyl, isopropyl, and t-butyl halides. Since the methyl group has a + I effect, the larger the number of methyl group on the carbon atom of the C—X the greater will be the electron density on this carbon. If we use one arrow-head to represent (qualitatively) the electron-repelling effect of a methyl group, then we have the following state of affairs.

$$\mathrm{Me} \rightarrow \mathrm{CH_2} \rightarrow \mathrm{X} \qquad \begin{matrix}\mathrm{Me} \searrow \\ \mathrm{Me} \nearrow\end{matrix} \mathrm{CH} \twoheadrightarrow \mathrm{X} \qquad \mathrm{Me} \rightarrow \begin{matrix}\mathrm{Me} \searrow \\ \\ \mathrm{Me} \nearrow\end{matrix} \mathrm{CH} \twoheadrightarrow \mathrm{X}$$

This increasing electron density on the central carbon atom increasingly opposes attack at this carbon atom by a negatively charged nucleophilic reagent.. Thus, the formation of the transition state for the S_N2 mechanism becomes increasingly difficult. It can therefore be anticipated that the S_N2 mechanism is made more difficult in passing from EtX to t-BuX. On the other hand, space the S_N1 mechanism involves ionisation as the first step, then it can be expected that as the electron density increases on the central carbon atom, the bonding pair in C—X becomes more and more displaced towards the X atom, and consequently ionisation of X as a negative ion will become easier. It therefore follows that the tendency for the SN1 mechanism should increase from EtX to t-BuX. These anticipated results have been realised in practice. Hughes, Ingold *et al*, (1935-1940) examined the hydrolysis of alkyl bromides in alkaline aqueous ethanol and showed that MeBr and EtBr undergo hydrolysis by the S_N2 mechanism, isoPrBr by *both* S_N2 and SN1 mechanism, and t-BuBr by the S_N1 mechanism only. It has also been shown, however, that the actual position where the mechanism changes over from S_N2 to SN1 in a given graded polar uses (such as the one above) is one fixed, but also depends on other factors such as the conformation of the nucleophilic reagent and the nature of the solvent.

An alternative explanation for the effect of the nature of R is as follows. When the S_N1 mechanism is favoured, it means that the activation energy for this path is less than that for the S_N2 and vice versa. Since the S_N1 reaction proceeds through a carbonium ion, the more stable this ion, the lower is the activation energy and the more favoured will be this mechanism. Now the order of stabilities of carbonium ions is p < s < t, and hence the tendency for the SN1 mechanism to operate will be t-BuX > iso rX > EtX, and the reverse tendency for the S_N2 mechanism.

Steric effects. The steric effect was originally thought to be a spatial effect brought into play be mechanical interference between groups, and was described as *steric hindrance.* The term steric hindrance considered the geometry of the *reactant* molecule. When however a molecule undergoes chemical reaction, it does so via a transition state. Consequently the geometry of both the initial and transition state must be taken into consideration. Thus steric forms may affect the speed and the mechanism of a reaction. It is however, often very difficult is distinguish between steric and polar factors. Nevertheless, the effects of the steric factor can often be assessed in some sort of qualitative manner. When steric effects slow down a reaction, the reaction is said to be subject to steric hindrance (*or* retardation), and when they speed up a reaction, that reaction is said to be subject to steric Halogen Derivatives of the Alkanes.

In an S_N2 reaction there will be five groups 'attached' to the carbon atom at which reaction groups. Thus there will be 'crowding' in the transition state, and the bulkier the groups, the greater will be the *compression energy,* and consequently the reaction will be hindered specially.

Let us consider the following example. Hughes *et al.* (1946) examined the rate of exchange of made ions in acetone solution with MeBr, EtBr, isoPrBr and t-BuBr under conditions where only the S_N2 mechanism operated. The relative reactivites were found to be 10,000; 65; 0-50;

$$\bar{I}\ \ R{-}Br \longrightarrow \overset{\delta-}{I}\cdots R\cdots\overset{\delta-}{Br} \longrightarrow I{-}R + Br^-$$

Thus, increasing the number of methyl groups on the central carbon atom increases the step retardation.

The problem is some what different for the SN1 mechanism Here the transition state do not contain more than four groups attached to the central carbon atom , and hence one word expect steric hindrance to be less important in the SN1 mechanism If, however, the molecular contains bulky groups, then by ionising, the molecule can relieve the steric strain, since carbonium ion-produced is flat (trigonal hybridisation) and so there is more room to accomodate the three alkyl groups. Thus, in such S_N1 reactions, there will be steric Halogen Derivatives of the Alkanes. Brown *et al* (1949) showed that the solvolysis of tertiary halides is subject to steric accelerate (*solvolysis is a* nucleophilic substitution reaction in which the *solvent* is the nucleophilic reagent

$$R_3C—X \xrightarrow{H_2O} R_3C^+—X^-$$

(tetrahedral; large strain) (trigonoal; planar; small strain)

These authors showed that as R increases in size, the rate of solvolysis increases. However, a larger the R groups are, the more slowly will the carbonium ion be expected to react with slowly molecules, since steric strain will be reintroduced into the molecule. In cases like this, the bonium ion tends to undergo an elimination reaction to form an alkene, and Brown *et al*. (1958) have shown that this elimination process increases as the alkyl groups become larger.

It should be noted that a fundamental part of the S_N1 mechanism is the postulate of the trances existence of carbonium ions. Triarylmethyl carbonium ions have been isolated as their salts, *e.g.*, phenylmethyl carbonium ion, Ph_3C^*, as its perchlorate and borofluoride (Dauben *et al*., 1960). The stability of ions such as these is attributed to resonance. The order of stability of a carbonium ions is t > s > p, and so one is more likely to expect success in the isolation of the Olah *et al*. (1964) have prepared, e.g, $Me_3C^*\ SbF_6^-$, and their infra-red spectra studies have to stantiated be planar sp^2 hybridised structure of simple alkyl carbonium ions.

Now let us consider the effect of the nature of the halogen atom. Experimental work be shown that the nature of X has very little effect, if any, on *mechanism,* but it does affect the rat of reaction for a given mechanism, *e.g.*, it has been found that in both S_N2 and S_N1 reactions the rate follows the order RI > RBr > RCI. Experimental work has shown that many S_N2 reactions of alkyl chlorides and bromides are catalysed by the presence of iodide ions. This is be explained on the basis that the activation energies of *both* steps in reaction (i) are lower that the activation energy of the one-step reaction (ii).

(i) $I^- + R—Cl \longrightarrow I—R + Cl^- \xrightarrow{OH^-} ROH + I^- + Cl^-$

(ii) $HO^- + R—Cl \longrightarrow HO—R + Cl^-$

The more pronounced the nucleophilic activity reagent, then more the S_N2 mechanism will be favoured, since in the S_N1 mechanism the reagent does not enter into the rate-determining step of ionisation. However, it can also be expected that as the nucleophilic that the mechanisms will change from S_N2 to S_N1. Hughes. Inglod , *et al.*, (1935) examined the mechanism and rate of decomposition of various trimethylsulphonium salts (Me_3S^+X) and showed that for X = OH, the mechanism is S_N2, and X = I, Cl or Br, the mechanism is S_N1.

Mechanism and rate of reaction are very affected by the nature of the solvent. It has been found that the ionising power of a solvent depends on its dielectric constant and its power of solvation, and it appears that the later is more important than the former. Solvation is due to the 'attachment' between solvent and solute molecules, and one important contributing factor attraction of a charge for a dipole. Since electrostatic work is done in the process of solvation, energy is lost by the system, and consequently the system is more stable. Furthermore, the greater the polarity (dipole moment) of the solvent, the greater is the attraction between solute and absent molecules and consequently the more stable is the system.

Thus, increasing the dielectric constant of the solvent increases the ionising potentiality of the solute molecules, and the higher the polarity of the solvent the greater is the amount of solvation. Solvation, however, may also be partly due to hydrogen bonding of the solvent (if this is protic) with the solute (if this can from hydrogen bonds). Some common solvents and their dielectric constants and dipole comments (D) are: water (81 : 1 ; 2– 3); formic acid (48.0; 1. 5), nitrobenzene (35.7, 4.0), ethanol (258, 1.7); acetone (21.3; 3.0), acetic acid (7.1, 1 – 1 : 5); chloroform (4.6; 1.1); ether (4.3;1.25); become (2.3; 0); carbon tetrachloride (2.2; 0).

When solute molecules have ionised, the oppositely charged pair of ions may become enclosed in a 'cage' of the surrounding solvent molecules, and may therefore recombine before they can escape from the cage. Such a complex is known as an *ion-pair,* and their recombination is known *is internal return.* It has now been shown that many organic reactions proceed via ion-pairs rather than dissociated ions. Thus we have the possibility of the following steps.

$$\underset{}{RX} \rightleftharpoons \underset{\text{ion – pair}}{R^+X^-} \rightleftharpoons \underset{\text{dissociated ions}}{R^+ \quad + X^-}$$

Many attempts have been made to correlate reaction rates and nature of the solvent. Hughes and Ingold (1935, 1948) proposed the following qualitative theory of solvent effects.

(i) Ions and polar molecules, when dissolved in polar solvents, tend to become solvated.

(ii) For a given solvent, solvation tends to increase with increasing magnitude of charge on the solute molecules or ions.

(iii) For a given solute, solvation tends to increase with the increasing dipole moment of the solvent.

(iv) For a given magnitude of charge solvation decreases as the charge is spread over a larger volume.

(v) The decrease of solvation due to the dispersal of charge will be less than that due to its deutuction.

Let us consider the following S_N1 reaction.

$$R—X \underset{+}{} \xrightleftharpoons{\text{slow}} \overset{\delta+}{R}...\overset{\delta-}{X} \xrightarrow{\text{fast}} R^+ + X^- \xrightarrow[\text{fast}]{OH^-} R—OH\cdot$$

Since the transition state has a larger charge than the reactant molecule, the former will be more solvated than the latter (rule ii). Thus the transition state is more stabilised that the reactant molecule, *i.e.*, the energy of activation is lowered, and so the reaction proceeds faster than had there been no (or less) solvation.

The rates of S_N2 reactions are also affected by the polarity of the solvent.

$$H\bar{O}\ R—X \rightleftharpoons \overset{\delta-}{HO}\cdots R\cdots\overset{\delta-}{X} \longrightarrow HO—R + X^-$$

A solvent with a high dipole moment will solvate both the reactant ion (nucleophilic reagent) and the transition state, but more so the former than the latter, since in the latter the charge, although unchanged in magnitude ($\delta - = -1/2$), is more dispersed than in the former. Thus fore solvation tends to stabilise the reactants more than the transition state (rule iv). Thus a activation energy is increased, and so the reaction is retarded.

The rate of the following S_N2 reaction (Menschutkin reaction 1890) is increased as the polarity of the solvent increases.

$$R_3\dot{N}\ R—X \rightleftharpoons \overset{\delta-}{R_3—N}\cdots R\cdots\overset{\delta-}{X'} \longrightarrow R_4N^+X^-$$

Here the charge on the transition state is greater than that on the reactants. Hence the transist state more solvated than the reactants and consequently stabilised, and so the activation is lowered.

The polarity of the solvent may *change* the mechanism of a reaction, *e.g.*, Olivier (1995) showed that the alkaline hydrolysis of benzyl chloride in 50 per cent, aqueous acetone process by both S_N2 and S_N1 mechanisms. When water was used as solvent, the mechanism was mainly S_N1. The dielectric constant of water is greater than that of aqueous acetone, an ionisation of the benzyl chloride is facilitated.

The above discussion has been purely qualitative, but it is also of interest to consider the problem a quantitative way. This involves the use of the following thermochemical cycle.

$$\begin{array}{ccc}
\text{RX (solvated)} & \xrightarrow[\text{(leassion)}]{1} & \text{R}^+ \text{ (solvated)} + \text{X}^- \text{ (solvated)} \\
\Big\downarrow 2 \text{ (desolvation)} & & 6 \Big\uparrow \text{(desolvation)} \quad 7 \Big\uparrow \text{(desolvation)} \\
 & & \text{R}^+ \text{ (vapour)} \quad \text{X}^- \text{ (vapour)} \\
 & & 6 \Big\uparrow \text{(desolvation)} \quad 7 \Big\uparrow \text{(desolvation)} \\
\text{RX (vapour)} & \xrightarrow[\text{(dessestion)}]{3} & \text{R}\cdot \text{ (vapour)} + \text{X}\cdot \text{ (vapour)}
\end{array}$$

The, if the enthalpy change for each step is represented by the number (of the step) as a subscript follows from Hess's law that

$$\Delta H_1 = \Delta H_2 + \Delta H_3 + \Delta H_4 + \Delta H_5 + \Delta H_6 + \Delta H_7$$

Where ΔH_4 is the ionisation potential of R. and ΔH_7 is the electron affinity of X.

All of these values except ΔH_2 have been calculated from suitable experimental data, and the rat of ΔH_2 has been estimated. The values given in the following discussion are those used by Frater is Singer (see References).

First let us consider the problem of reactions in the gas phase and in solution. In the gas phase, and reaction is homolytic, then ΔH for step 3 in the above cycle is ΔH_3. If we use methyl chloride and example, than ΔH_3 is + 334.7 kJ. If however, the reaction is heterolytic, then

$$\Delta H = \Delta H_3 + \Delta H_4 + \Delta H_5 = +\,334.7 + 970.7 - 364.0 = 941.4 \text{ kJ}$$

It can be seen that heterolytic reaction in the gas phase is very unfavourable energetically. If the reaction is carried out in water, then

$$\begin{aligned}\Delta H_1 &= +\,33.47 + 334.7 + 970.7 - 364.0 - 343.1 - 288.7 \\ &= 343.1 \text{ kJ}\end{aligned}$$

This much more favourable value is due to the large contribution of the heats of solvation (ΔH_6 , ΔH_7).

As we have already seen, in the S_N1 mechanism (with all ions and molecules solvate).

$$RX_{(aq)} \overset{slow}{\rightleftharpoons} R^+_{(aq)} + X^-_{(aq)} \xrightarrow[fast]{H_2O} ROH_{(aq)} + H^+_{(aq)}$$

The rate order is t-Bu > isoPr > Et > Me. The following Table gives the values of ΔH_1 the ionisation in kJ mol^{-1}) for various alkyl halides.

Group	Cl	.Br	I
Me	343.1	343.1	364.0
Et	238.5	213.4	238.5
isoPr	104.6	96.23	115.1
t-Bu	71.13	62.76	75.31

It can be seen in going in a downward direction for a given halogen atom, ΔH_1 decreases rapidly. Moreover, calculations have shown that ΔH_1 and E (activation energy) are almost identical. Hence the rate of the S_N1 reaction will increase in the downward direction *i.e.*, t-Bu >> isoPr > Me.

It has been shown experimentally that the rate of the S_N1 reaction (and that of the S_N2) for a given alkyl group is RI > RBr > RCI. The above Table indicates the order RBr > RCl > Rl. This again illustrates the point that even though a quantitative approach to a problem is always desirable, unless all the necessary values are known (reasonably) accurately, estimates may lead to some, wrong conclusions. This is more the case with reactions in solution than with gas-phase reactions.

2. Alkyl halides are reduced by dissolving metals to alkanes:

$$RX \xrightarrow{eH^+} RH + HX$$

Primary and secondary alkyl chloride and bromides are nor affected in natural of acidic media, whereas prim., s-and t-iodides, t-chlorides and t-bromides are converted into alkanes by *catalytic hydrpgenolysis, i.e.*, hydrogenation in the presence of a catalyst. The best catalyst is Pd—C, but Raney nickel is also effective provided it is used in large amounts.

$$RX + H_2 \xrightarrow{Pd-C} RH + HX$$

Reduction to alkane can also be carried out with lithium aluminium hydride or sodium borohydride.

Johnson *et al.* (1964) have shown that primary and secondary alkyl iodides are oxidised by dimethyl sulphoxide to aldehydes or ketones, respectively. The mechanism is uncertain, but a possibility is:

$$RCH_2I + Me_2{=}O \longrightarrow I^- + R{-}\underset{H}{CH}{-}O{-}SMe_2 \longrightarrow RCHO + Me_2S + H^+$$

3. Alkyl halides undergo the Wurtz reaction to form alkanes. On the other hand alkyl halides, when treated with certain metallic alloys, form *organometallic* compounds, *e.g.*, ethyl chloride heated with a sodium lead alloy under pressure gives tetraethyl-lead.

$$4C_2H_5Cl + 4Na/Pb \longrightarrow (C_2H_5)_4Pb + 4NaCl + 3Pb.$$

4. When an alkyl halide is heated at about 300°C, and at a lower temperature in the presence of aluminium chloride as catalyst, the alkyl halide undergoes rearrangement, *e.g.*,

$$CH_3CH_2CH_2Cl \xrightarrow{-Cl^-} CH_3CH_2CH_2^+ \longrightarrow CH_3\overset{+}{C}HCH_3$$
$$\xrightarrow{Cl^-} CH_3CHClCH_3.$$

If there is no hydrogen atom on the carbon adjacent to the C—X group, an alkyl group migrates, *e.g.*, neopentyl chloride rearranges to 2-chloro-2-methylbutane.

$$(CH_3)_3CCH_2Cl \longrightarrow (CH_3)_2CClCH_2CH_3$$

5. When alkyl halides are boiled with ethanolic potassium hydroxide, alkenes are obtained, *e.g.*, propyl bromide gives propene.

$$CH_3CH_2CH_2Br + KOH \xrightarrow{\text{ethanol}} CH_3CH = CH_2 + KBr + H_2O$$

This is an example of the 1,2- or β-elimination reactions, in which two groups are lost from *neighbouring* atoms to form a double bond. Other elimination types (x, g, etc) are also known but these are much less common. There are two important mechanism which operate with alkyl halides, E1 and E2, both of which are *polar* (ef, cyclic eliminations.

E1 mechanism (unimolecular elimination). This mechanism is parallel to the S_N1 mechanism in that the rate-determining step is ionisation to a carbonium ion, but the second step is now elimination of a β-proton to form an alkene.

$$HCR_2—CR_2—Z \xrightleftharpoons{\text{slow}} HCR_2—CR_2^+ + Z^-$$

$$B\ H—CR_2—CR_2 \xrightleftharpoons{\text{fast}} BH + CR_2{=}CR_2$$

The reaction is carried out in the presence of a base and the proton in transferred to this , *e.g.*, t-butyl bromide in aqueous ethanolic potassium hydroxide gives isobutene.

$$(CH_3)_3CBr \xrightleftharpoons{\text{slow}} (CH_3)_3C^+ + Br^-$$

$$HO^-\ H—CH_2—C(CH_3)_2 \xrightarrow{\text{fast}} H_2O + CH_2{=}C(CH_3)_2$$

Evidence for the E1 mechanism has been obtained in several ways, *e.g.*, the rate low for many eliminations is first order (alkyl halide only), *i.e.*, rate = k [R.X]. This is consistent with the mechanism given.

E2 mechanism (bimolecular elimination). The general reaction may be written.

$$B\ H—CR_2—CR_2—Z \longrightarrow BH—CR_2{=}CR_2 + Z^-$$

An example of this type is the action of ethanolic potassium hydroxide on isopropyl bromide.

$$Et\bar{O}\ H—CR_2—CH(CR_3)—Br \longrightarrow EtOH + CH_2{=}CHCH_3 + Br^-$$

Evidence for the E2 mechanism comes from the fact that the rate law for may eliminations is seconds order, *i.e.*, rate = k [RX] [B]. There is, however, a difficulty here. The second-order rate law is consistent with the one-step mechanism given above, but it is also consistent with the following two-step mechanism (Ph = C_6H_5).

$$EtO + PhCH_2CH_2Br \xrightleftharpoons{\text{fast}} EtOH + PhCHCH_2Br$$

$$Ph\bar{C}H—CH_2—Br \xrightarrow{\text{slow}} PhCH{=}CH_2 + Br^-$$

A simplified derivation of the rate law for this reaction is as follows. Since the reaction is carried out in ethanol, the concentration of the *ethanol produced* in the first step may be neglected. Hence.

$$K = [Ph\bar{C}HCH_2Br]/([PhCH_2CH_2Br][OEt^-])$$

$$\text{rate} = k = [Ph\bar{C}HCH_2Br] = kK\ [PhCH_2CH_2Br][OEt^-].$$

Since k and K are both constants, their product may be replaced by a constant, and the rate law

$$\text{rate} = k\ [PhCH_2CH_2Br]\ [OEt^-]$$

i.e., the rate law is second-order. Thus, kinetics cannot distinguish between these two possible elimination mechanisms.

In the second mechanism, a slow unimolecular elimination occurs in the conjugate base (cB) of the reactant, and hence this mechanism is called the E1 cB or carbanion mechanism. Since the first step must be reversible (acid-conjugate base equilibrium), if ethanol containing EtOD is used as solvent,

would be expected that the original bromide would incorporate deuterium. Skell at al. (1945) examined this reaction and found that after the reaction and found that after the reaction was half-completed, the recovered bromide did not contain deuterium. Hence *this* mechanism is not E1cB, but is E2.

It appears that definite examples of the E1cB mechanism are rate.

Which mechanism, E1 or E2 operates depends on the nature of the alkyl halides and the conditions used. Before discussing this, let us consider the situation where two alkanes may be formed by dehydrohalogenation of an alkyl halide. Originally, the direction of elimination was described empirically by Saytzeff's rule (1875). This rule may be sated in two ways.

The predominant product is the most substituted alkene, *i.e.*, the one carrying the largest number of alkyl substituents; or Hydrogen is eliminated preferentially from the carbon atom joined to the heat number of hydrogen atoms. There is also another rule which describes empirically the direction elimination in quaternary ammonium hydroxide. This rule, the Hofmann rule (1881), is discussed later, but may be sated here as follows. The predominant alkene (from an onium hydroxide) is the least substituted alkene. An illustration of these rules is the dehydro bromination of 2-bromobutane.

$$\underset{\text{(I)}}{CH_3CH=CHCH_3} \longleftarrow CH_3CH_2CHBrCH_3 \longrightarrow CH_3CH_2CH=CH_2$$

(I) is the Saytzeff product and (II) is the Hofmann product, and since we are dealing with elimination from an alkyl halide, (I) will be expected to be the predominant product,. Ingold *et al.* (1994) have offered the following explanation for the Saytzeff rule. If (I) and (II) are produced by an E2 mechanism, their formation may be written as follows.

$$\text{(i)}\ HO^- \ H—\underset{CH_3}{CH}—\overset{CH_3}{CH}—Br \longrightarrow H_2O + \underset{CH_3}{CH}=\overset{CH_3}{CH} + Br^- \quad \text{(I)}$$

$$\text{(ii)}\ HO^- \ \overset{H—CH_2}{CH_2}—\underset{CH_3}{CH}—Br \longrightarrow H_2O + CH_2—\overset{CH_2}{\underset{CH_3}{CH}} + Br^- \quad \text{(II)}$$

In the transition state of (i), the CH_3 group can enter into hyperconjugation with the partly formed double bond, there by lowering the energy of the transition state. In (ii), however, the CH_3 group cannot enter into hyperconjugation with the partly formed double bond, and so the energy of this transition state is higher than the of (i), and consequently the latter path is favoured.

On the basis that the hyperconjugative effect determines the stability of the alkene produced, it can be deduced that the ease of formation of alkene form an alkyl halide by the E2 mechanism should be t—R > s—R > p—R, *e.g.*, t—Bu, (2 Me groups) > isoPr (1 Me group) > Et (No Me groups.

$$\begin{array}{ccc} H—\overset{H}{C}\cdots H\cdots \overset{\delta-}{O}Et & H—\overset{H}{C}\cdots H\cdots \overset{\delta-}{O}Et & H—\overset{H}{C}\cdots H\cdots \overset{\delta-}{O}Et \\ Me—\underset{Me}{C}\cdots Br^{\delta-} & Me—\underset{H}{C}\cdots Br^{\delta-} & H—\underset{H}{C}\cdots Br^{\delta-} \end{array}$$

These predictions are borne out in practice. Also, since the most substituted alkene is general the most stable isomer, it therefore follows that Saytzeff's rule leads to the predominant of the most stable product. A further point in this connection is that when the alkene is capable of existing as geometrical isomers, the *trans*-isomer generally predominates over the *cis*.

Saytzeff's rule also applies when the mechanism is E1. In this mechanism the formation of a carbonium ion is the rate-determining step (as for the S_N1 mechanism), and then the carbonium eliminates a proton to form an alkene, the stability of which will be largely determined by hyperconjugative effect.

Stereochemistry of elimination reactions. Many examples in the literature show that *trans* elimination occurs more readily than *cis*., chlorifumaric acid undergoes dehydrochlorinating about fifty times as fast times as fast as chloromaleic acid.

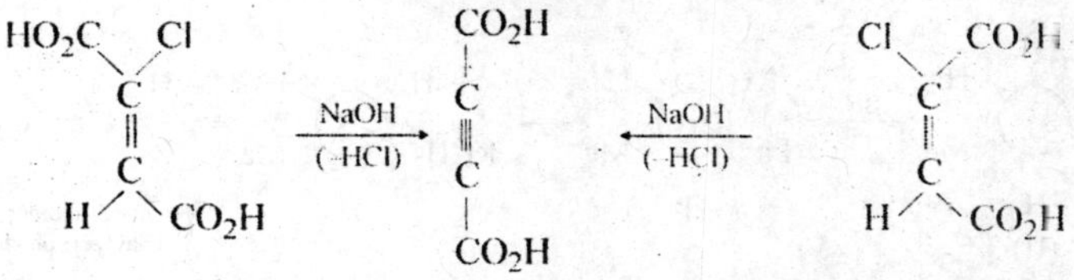

$$HO_2C(Cl)C{=}C(H)CO_2H \xrightarrow[(-HCl)]{NaOH} HO_2C{-}C{\equiv}C{-}CO_2H \xleftarrow[(-HCl)]{NaOH} Cl(CO_2H)C{=}C(H)CO_2H$$

Experiments of this type (involving 'rigid' systems) let Hughes and Ingold to propose the bimolecular elimination reactions take place when the two groups to be eliminated are *trans* and lie in one plane with the two carbon atoms to which they are attached, *i.e.*, E2 reactions an stereoselectively *trans*.

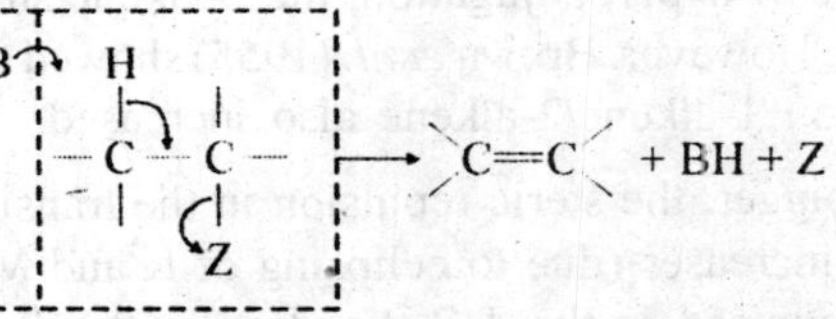

$$B^{\frown}\; H{-}\overset{|}{C}{-}\overset{|}{C}{-}Z \longrightarrow \;>C{=}C<\; + BH + Z$$

Evidence for E2 *trans*-elimination has also been obtained from acyclic systems, *e.g.*, Crist et. al. (1952) showed that the alkenes produced by the base-catalysed dehydrpbromination of the diastereoisomeric 1-bromo-1,2-diphenylpropanes (I and (II) can only arise by *trans*-eliminates.

Later work has shown that bimolecular eliminations may, in certain circumstances, proceed by *cis*-elimination.

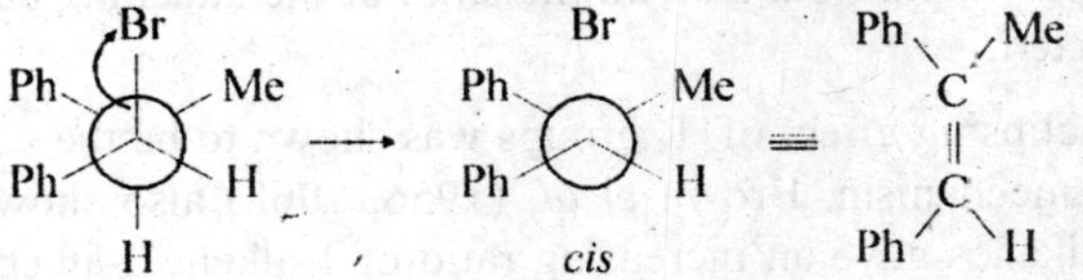

(I): Br, Ph, Me, Ph, H, H → Ph, Me, Ph, H (*cis*) ≡ Ph(Me)C=C(Ph)H

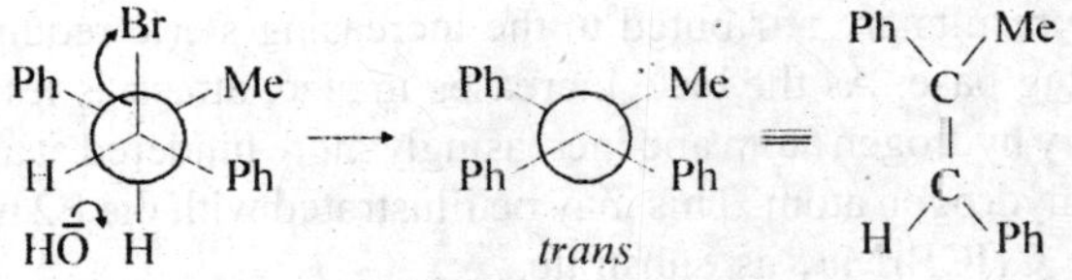

(II): Br, Ph, Me, H, Ph, HŌ H → Ph, Me, Ph, Ph (*trans*) ≡ Ph(Me)C=C(H)Ph

Steric Effects in Elimination

Let us consider the E1 reaction with RCH_2CMe_2Br. Accepting that elimination of HBr is selectively *trans*, then we may represent the two directions of elimination as follows.

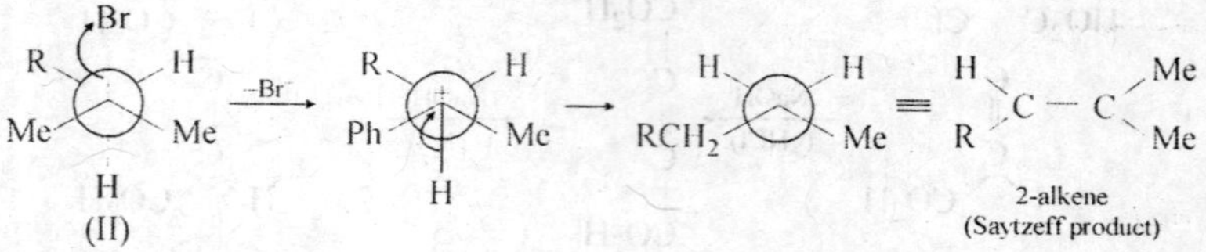

(II) $\xrightarrow{-Br^-}$ $\longrightarrow$ $\equiv$ 2-alkene (Saytzeff product)

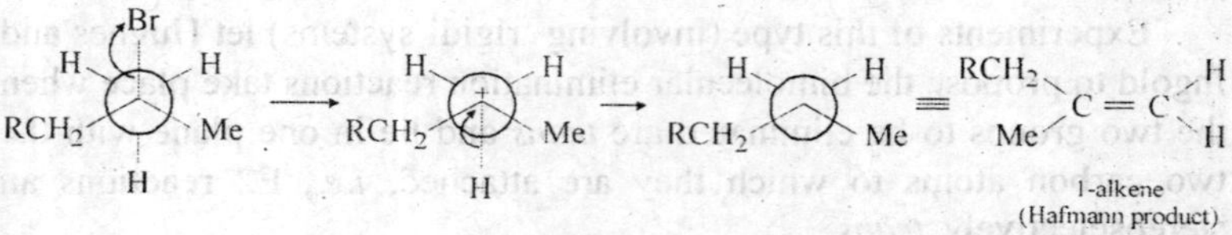

$\longrightarrow$ $\longrightarrow$ $\equiv$ 1-alkene (Hafmann product)

On the basis of hyperconjugation, the 2-alkene should predominate (Saytzeff's rule). However, Brown *et al.* (1955) showed that as R increased in size, the ration 1-alkene/2-alkene also increased.

As R gets bigger, the steric repulsion in the transition state leading to the 2-alkene increases (due to eclipsing of R and M), whereas steric repulsion is minimised in the T.S. Leading to the 1-alkene (Hofmann rule).

Thus, in this case, the 1-alkene is more stable than the 2-isomer. Hence, in the E1 mechanism, the structure of the alkene is determined by hyperconjugation and steric effects.

The E2 mechanism is more complicated, since not only does the size of R play a part, but also do the sizes of the attacking base and the group ejected.

The eclipsing effect of R groups was shown to be the same as that in the E1 mechanism. Brown *et al.* (1956, 1968) also showed that 2-halogenoalkanes gave an increasing ratio of 1-alkene.2-alkene with the potassium alloxides EtO^-K^+,K^+, $Me_3CO^-K^+$.

These results are attributed to the increasing steric requirements of the attacking base..As the base increases in size, attack is less hindered co a primary hydrogen atom and increasingly more hindered at a secondary or tertiary hydrogen atom. This may be illustrated with the E2 mechanism using $Me_2CHCBrMe_2$ as substrate.

(III) Newman projection: $RO^- \curvearrowright H$; Me, Me (front); Me, Me (back); Br $\longrightarrow$ $Me_2C{=}CMe_2$ 2-alkene (Saytaff product)

(IV) Newman projection: $RO^- \curvearrowright H$; Me, $CHMe_2$; Me, H; Br $\longrightarrow$ $Me_2CH(Me)C{=}CH_2$ 1-alkene (Hofmann product)

As R becomes larger, the steric effect increase more in (III) than in (IV) RO^- approaches from the front), and so 1-alkene formation increases.

It might be noted here that this change in direction of elmination may be used to shift a double bond in an alkene, *e.g.*,

$$Me_2C = CMe_2 \xrightarrow{HCl} Me_2CHCMe_2Cl \xrightarrow{Et_5CO^-K^-}$$

$$Me_2CMCMe - CH_2 \quad (97\%)$$

Brown *et al.* (1956) showed that as the size of the leaving group increases, the yield of 1-alkene also increases (for a given base). Here again, there is less steric hindrance in (IV) than in (III) as Z gets larger (replace Br by Z).

Now let us consider the conditions which favour the formation of alkenes from alkyl halides. In practice, the situation is complicated by the fact that substitution and elimination often occur together.. However, it has been shown that under $E1/S_N1$ conditions, substitution is more favoured than under E2/SN2 conditions. Thus, for a given alkyl halide, the alkene yield is higher for the E2 mechanism than for the E1. Since a high concentration of the base with high nucleophilic power favours the E2 mechanism, alkene preparation is best carried out in solvents with ionising power. Thus, ethanolic sodium, hydroxide solution (solvent of low ionising power and a very strong base, OEt^-) gives better alkene yield than does aqueous sodium hydroxide (solvent of high ionising power and a base. OH^-, which is weaker than OEt^-.). Also, a base of and nucleophilic power will have little affinity for hydrogen, and consequently alkene formation will be decreased and substitution encouraged. The acetate ion a very weak base, and hence its use will given an ester as the main product rather than alkene. Thus, the conversion of an alkyl halide into the alcohol is often best carried out via the ester, which is then hydrolysed.

A point of interest here is that even under the most favourable conditions, ethyl halides give only 1 per cent, of ethylene, the main reaction leads to substitution products.

The eliminations discussed above have dealt with only alkyl halides as substrates.

However, many other types of alkyl compounds also undergo E1 and E2 reactions, *e.g.*, in RZ, Z may be R_3N^+ R_2S^+, R_3P^+, OTs, OCOR.etc., and the base may be OH^-, OR^-, NH_2^-, etc.

6. When alkyl halides are heated with ethanolic ammonia under pressure, a mixture of *ammonias i.e.*, substituted ammonias, is obtained, *e.g.*, .

$$RX + NH_3 \longrightarrow RNH_2 + HX \longrightarrow RNH_2.HX \text{ or } [RNH_3]^+X^-$$

7. When alkyl halides are heated with aqueous ethanolic potassium cyanide, alkyl *cyanides* are obtained.

$$RX + KCN \longrightarrow RCN + KX$$

When aqueous ethanolic silver cyanide is used, cyanides and *isocyanides* are formed.

$$RX + AgCN \longrightarrow RCN + AgX$$

$$RX + AgCN \longrightarrow RCN + AgX$$

8. Alkyl halides form esters when heated with an ethanolic solution of a silver salt.

$$R^1CO_2Ag + R^2X \longrightarrow R^1CO_2R^2 + AgX$$

The acid may be organic or inorganic. If the salt is silver nitrite two isomeric compounds are obtained, the *nitrite* (ester) and the *nitro-* compound.

$$R—O—N{=}O \longrightarrow RX + Ag—O—N{=}O \longrightarrow R—\overset{+}{N}\begin{smallmatrix}{/\!\!/}O\\ \searrow O^-\end{smallmatrix}$$

9. When alkyl halides are heated with sodium alkoxides, *ethers* are obtained.

$$R^1ONa + R^2X \longrightarrow R^1OR^2 + NaX$$

10. Alkyl halides heated with aqueous ethanolic sodium hydrogen sulphide form *thioalcohols* and *thioethers,* when mercaptides are used.

$$RSH \xleftarrow{NaSH} RX \xrightarrow{RSNa} RSR$$

With sodium sulphite, a *sulphonate* is formed.

$$RX + Na_2SO_3 \longrightarrow RSO_3Na + NaX$$

11. Alkyl halides may be used in the *Freidel-Crafts* reaction and to prepare Grignard reagents.

Methyl chloride is prepared industrially by chlorinating methane with chlorine diluted with nitrogen, the ration by volume of CH_4:Cl_2:N_2 being 8:1:80. The reactions is carried out in the presence of partly reduced cupric chloride as catalyst. All four chloromethanes are obtained, the methyl chloride comprising 90 per cent of the chlorine used.

$$CH_4 + Cl_2 + (N_2) \longrightarrow CH_3Cl + HCl + (N_2) + (CH_2Cl_2 + CHCl_3 + CCl_4)$$

By adjusting the ration of chlorine to methane, each chloromethane can be obtained as the main product.

Methyl chloride, b.p – 24°C, is used in the manufacture of aniline dyes, silicone rubbers, tetramethyl, methylation of cellulose, as a refrigerating agent, local anesthetic and as a fire extinguisher.

Methyl bromide, b.p. 45°C, is manufactured by passing the vapours of methanol and hydrobromic acid iodide is prepared industrially.

Methyl iodide is prepared industrially.

(i) By warming a mixture of methanol and red phosphorus with iodine.

$$6CH_3OH + 2P + 3I_3 \longrightarrow 6CH_3I + 2H_3PO_3$$

(ii) By the action of methyl sulphate on potassium iodide solution in the presence of calcium carbonate.

$$KI + (CH_3)_2SO_4 \longrightarrow CH_3I + K\,(CH_3)SO_4 \quad (90\text{–}94\%)$$

Methyl iodide is a sweet-smelling liquid. b.p. 42.5°C. Since it is liquid, it is easier to handle than methyl chloride, and so is used a great deal as a methylating agent in laboratory organic syntheses, but the chloride is used industrially because it is cheaper.

Ethyl chloride is prepared industrially.

(i) By the chlorination of ethane at about 400°C.

$$C_2H_6 + Cl_2 \longrightarrow C_2H_5Cl + HCl$$

(ii) By the addition of hydrogen chloride to ethylene (from cracked petroleum) in the presence of aluminium chloride as catalyst.

$$C_2H_4 + HCl \xrightarrow{ACl_3} C_2H_5Cl$$

Ethyl chloride, b.p. 12.5°C, is used in the preparation of tetraethyl lead, etc, and as a refrigerating agent.

Ethyl bromide, b.p. 38.5°C, is manufactured by irradiating a mixture of ethylene and hydrogen bromide with γ-radiation from a cobalt—60 source.

$$HBr \xrightarrow{\gamma-rays} H^{\cdot} + Br^{\cdot}$$

$$C_2H_4 + Br^{\cdot} \longrightarrow {}^{\cdot}CH_2 - CH_2Br \xrightarrow{HBr} CH_3CH_2Br + Br^{\cdot}$$

It is used for, preparing barbiturate drugs.

DIHALOGEN DERIVATIVES

General Methods of Preparation

1,1-Dihalides may be prepared

(i) By the action of phosphours pentahalides on aldehydes or ketones, *e.g.*, acetone gives isopropylidene dichloride when treated with phosphorus pentachloride.

$$CH_3COCH_3 + PCl_5 \longrightarrow CH_3CCl_2CH_3 + POCl_3$$

(ii) By the addition of halogen acids to alkynes, *e.g.*, ethylidene dibromide from acetylene and hydrogen bromide.

$$CH \equiv CH + HBr \longrightarrow CH_2 = CHBr \xrightarrow{HBr} CH_3CHBr_2$$

1,2,-Dihalides may be prepared by the addition of halogen to alkenes, *e.g.*, propene dibromide from propene and bromine.

$$CH_3CH{=}CH_2 + Br_2 \longrightarrow CH_3CHBrCH_2Br$$

The method of preparing an αω-dihalide is special to the particular halide, *e.g.*, trimethylene dibromide may be prepared by the additicn of hydrogen bromide to alkyl bromide at low temperatures.

$$CH_2{=}CHCH_2Br + HBr \longrightarrow CH_2BrCH_2CH_2Br$$

Oldham (1950), however, has found that αω-dibromides may be prepared by the action of bromine in carbon tetrachloride on the silver salt of a dicarboxylic acid.

$$(CH_2)_n(CO_2Ag)_2 + 2Br \longrightarrow Br(CH_2)_nBr + 2AgBr + 2CO_2$$

The yields are very good provided n is 5 or more.

GENERAL PROPERTIES AND REACTIONS OF THE DIHALIDES

The dihalides are sweet-smelling, colourless liquids.

1,1,-Dihalides are hydrolysed by aqueous alkalis to the correspond... carbonyl compound (aldehyde or ketone), *e.g.*, isopropylidene dichloride gives acetone.

$$CH_2CCl_2CH_3 + 2KOH \longrightarrow 2KCl + [CH_3C(OH)_2CH_3]$$
$$\longrightarrow CH_3COCH_3 + H_2O$$

It has generally been found that a compound containing two (or more) hydroxyl groups attached to the *same* carbon atom is unstable, and readily eliminates a molecule of water,

1,1,-Dihalides gives alkenes when treated with zinc dust and methanol, and alkynes when treated with ethanolic potassium hydroxide.

1,2- and αω–Dihalides are just as reactive as the alkyl halides. When heated with zinc and methanol, 1,2-dihalides give alkenes, but α,ω-dihalides in which the two halogen atoms are in the 1.3, to the 1,6 positions given cyclic compounds, *e.g.*, trimethylene dibromide gives cyclopropane.

$$CH_2\begin{cases}CH_2Br\\CH_2Br\end{cases} + Zn \longrightarrow CH_2\begin{cases}CH_2\\|\\CH_2\end{cases} + ZnBr_2$$

1,2-Di-iodides tend to eliminate iodine, particularly at raised temperatures, to form alkenes *e.g.*, propene di-iodide gives propene.

$$CH_3CHICH_2I \xrightarrow{\text{heat}} CH_3CH = CH_2 + I_2$$

Methylene dichloride, b.p., 40°C, is prepared industrially by the direct chlorination of methane (see methyl chloride). It is used as an industrial solvent. Methylene dibromide, b.p., 97°C, and methylene di-iodide, b.p. 181°C, are prepared by the partial reduction of bromoform and iodoform, respectively, with sodium arsenite in alkaline solution (the yield of CH_2Br_2 is 88–90 per cent, CH_2I_2, 90–97 per cent), *e.g.*,

$$CHI_3 + Na_3AsO_3 + NaOH \longrightarrow CH_2I_2 + NaI + Na_2AsO_4$$

All the methylene dihalides are used in organic syntheses.

Ethylene dichloride, b.p. 84°C, and ethylidene dichloride, b.p. 57°C, are isomers, the former is prepared from ethylene and chlorine, and the latter by the action of phosphorus pentachloride on acetaldehyde (q.v.).

Ethylene dibromide, b.p. 131°C, is made industrially by passing ethylene into cooled bromine, it is used as a component in anti-knock fluids.

TRIHALOGEN DERIVATIVES

The most important derivatives are those of methane, and they are usually know by their trivial names, chloroform, $CHCl_3$, bromoform, $CHBr_3$ and iodoform, CHI_3.

Coloform may be prepared in the laboratory of industrially by heating ethanol or acetone with bleaching power, or with chlorine and alkali (yield about 40 per cent). The reactions is externally complicated, and the mechanism is obscure.

The equations usually given for the action of bleaching power on ethanol are, (i) oxidation of ethanol to acetaldehyde, (ii) chlorination of acetaldehyde to trichloroacetaldehyde, (iii) decomposition of trichloroacetaldehyde (choral) by free calcium hydroxide (present, in the bleaching power) into chloroform and formic acid.

(i) $CH_3CH_2OH + Cl_2 \longrightarrow CH_3CHO + 2HCl$

(ii) $CH_3CHO + 3Cl_2 \longrightarrow CCl3CHO + 3HCl$

(ii) $2CCl_3CHO + Ca(OH)_2 \longrightarrow 2CHCl_3 + (HCO_2)_2Ca$

When acetone is used, the first product given is trichloroacetone, which is then decomposed by the calcium hydroxide into chloroform and acetic acid.

(i) $CH_3COCH_3 + Cl_2 \longrightarrow CCl_3COCH_3 + 3HCl$

(ii) $2CCl_3COCH_3 + Ca(OH)_2 \longrightarrow 2CHCl_3 + (CH_3CO_2)_2Ca$

Chloroform is also prepared industrially by the chlorination of methane (see methyl chloride).

Chloroform is a sickly, sweet-smelling, colourless liquid, b.p. 61°C. It is sparingly soluble is water but readily soluble in ethanol and ether. It does not burn in air under usual conditions but its vapour may be ignited, when it burns with a green-edged flame.. According to Hine (1953 1954), chloroform (and other haloforms) undergoes alkaline hydrolysis to produce the formation and carbon monoxide by what Hine calls the *alpha-elimination mechanism.* This involves the removal of hydrogen and chloride ions form the *same* carbon atom.

The mechanism proposed is an follows, involving the intermediate formation of dichloromethylene

$$CHCl_3 + OH^- \rightleftharpoons CCl_3^- + H_2O$$

$$:CCl_3^- \longrightarrow :CCl_2 + Cl^-$$

$$:CCl_2 \xrightarrow[H_2O]{OH^-} CO + HCO_2^- + 3Cl^-$$

The removal of a proton is facilitated by the – 1 effect of the chlorine atoms. However, the cannot be the complete story since, although fluorine has a much stronger – 1 effect than chlorine, $CHClF_2$ behaves differently. Hine *et al.* (1954 – 57) have shown that $CHCl_3$, $CHBr_2$ $CHBr_2F$, etc, readily undergo deuterium exchange in alkaline solution.

$$CHCl_3 + OH^- \rightleftharpoons H_2O + :\bar{C}Cl_3 \overset{D_2O}{\rightleftharpoons} CDCl_3 + OD^-$$

$CHClF_2$ was found to react more slowly. To account for this, it has been suggested that the CCl_3^- ion is stabilised by resonance. Cl can expand its shell to a decet because it has *d*-orbitals but F has no*d*-orbitals and so this expansion cannot take place.

$$:\ddot{C}l{-}\underset{|\atop :\ddot{C}l:}{\overset{:\ddot{C}l:\atop |}{C}}{:}^- \longleftrightarrow :\ddot{C}l{-}\underset{\|\atop :\ddot{C}l:}{\overset{:\ddot{C}l:^-\atop \|}{C}} \longleftrightarrow \cdots$$

This type or resonance, which requires expansion of an octet, is known as *d* **orbital resonance**.

When chloroform is treated with zinc and hydrogen chloride in ethanolic solution, methylene dichloride (q.v) is obtained, when treated with zinc and water, methane is obtained.

$$CHCl_3 \xrightarrow{Zn/H_2O} CH_4 + 3HCl$$

When chloroform is warmed with silver power, acetylene is obtained.

$$2CHCl_3 + 6Ag \longrightarrow C_2H_2 + 6AgCl$$

When treated with concentrated nitric acid, chloroform forms *chloropicrin.*

$$CHCl_3 + HNO_3 \longrightarrow CCl_3NO_2 + H_2O$$

Chloropicrin or nitrochloroform (liquid, b.p., 112°C) is used as an insecticide, and has been used as a war-gas. Chloroform adds on to the carbonyl group of ketones in the presence of potassium hydroxide, *e.g.*, with acetone it forms *chloretone* (colourless needles, m.p. 97°C), which is used as a drug. The mechanism is possibly.

$$(CH_3)_2C{=}O \; + \; :\bar{C}Cl_3 \longrightarrow (CH_3)_2C(O^-)CCl_3 \xrightarrow{H_2O} (CH_3)_2C(OH)CCl_3$$

POLYHALOGEN COMPOUNDS

The organic compounds containing more than one halogen atoms in their molecules are termed *compounds.* These compounds are designated as follows.

The compounds containing two halogen atoms in their molecules are called *dihalo*-compounds. Those containing three halogen atoms in their molecules are called *trihalo*- compounds. The compounds containing four halogen atoms in their molecules are called *tetrahalo*-compounds.

Some of the polyhalo compounds are commerically very important. Some common polyhalogen compounds are listed in Table 17.3 given on the next page.

DICHLOROMETHANE (METHYLENE CHLORIDE, CH_2Cl_2)

Dihalogen derivatives of alkanes are obtained by replacing its two hydrogen atoms by two halogen atoms. Dichloromethane is thus a derivative of methane in which two H-atoms of methane molecule are replaced by two Cl atoms.

$$H-\underset{H\;\;Cl}{C}-Cl$$

Dichloromethane

Industrial Preparation

Dichloromethane is prepared on industrial scale by direct chlorination of methane (or natural gas).

$$\underset{\text{methane}}{CH_4} \xrightarrow[-\,HCl]{+\,Cl_2\,,\,h\nu} \underset{\text{chloromethane}}{CH_3Cl} \xrightarrow[-\,HCl]{+\,Cl_2\,,\,h\nu} \underset{\text{dichloromethane}}{CH_2Cl_2}$$

$$\xrightarrow[-\,HCl]{+\,Cl_2\,,\,h\nu} \underset{\text{trichloromethane}}{CHCl_3} \xrightarrow[-\,HCl]{+\,Cl_2\,,\,h\nu} \underset{\text{tetrachloromethane}}{CCl_4}$$

In this reaction, a mixture of different chloro derivatives is obtained. Dichloromethane is separated from this mixture by fractional distillation.

Properties :

(i) Dichloromethane is a colourless, sweet-smelling, volatile liquid (b.p. 40°C). (ii) It has low inflammability.

Use : Dichloromethane is used

(i) as an effective extraction solvent in pharmaceutical and food industries.

(ii) as a paint remover.

Formula	*IUPAC name*	*Common name*
CH_2Br–CH_2Br (i.e. CH_2Br | CH_2Br)	**Dihalo compounds**	
$CHBr_2$–CH_3	1,2-dibromoethane	Ethylene bromide
CH_2—CH_2—CH_2—CH_2 (Br on first and last C)	1,1-dibromoethane	Ethyliden bromide
Benzene ring with Cl, Cl	1,4-dibromobutane	Tetramethylene bromide
Benzene ring with Cl, Cl	1,2-dichlorobenzene	*o*-dichlorobenzene

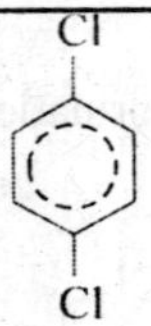

Cl

|

(benzene ring)

|

Cl

1,3-dichlorobenzene *m*-dichlorobenzene

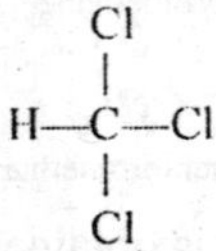

Cl
|
H—C—Cl
|
Cl

1,4-dichlorobenzene *p*-dichlorobenzene

Trihalo compounds

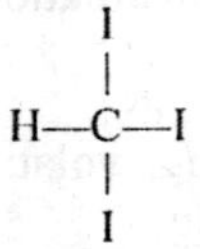

I
|
H—C—I
|
I

Trichloromethane Chloroform

Cl
|
Cl—C—Cl
|
Cl

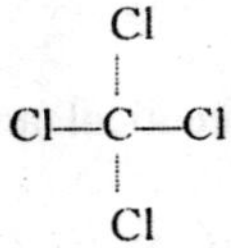

Triiodomethane Iodoform

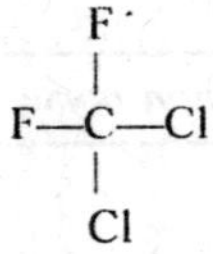

F
|
F—C—Cl
|
Cl

dichlorodifluoromethane Freon

Polyhalo compounds

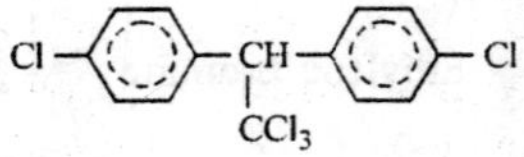

Cl—(benzene ring)—CH—(benzene ring)—Cl
|
CCl_3

2,2-bis (*p*-chlorophenyl)-1,1,

1-trichloroethane D.D.T.

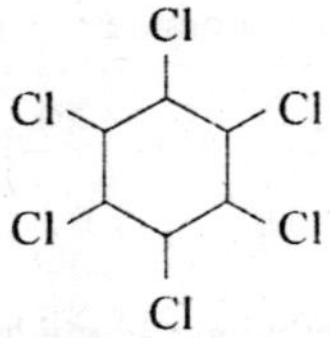

Cl, Cl, Cl, Cl, Cl, Cl (cyclohexane ring)

1,23,4,5,6-hexachlor- Benzenehexachloride

ocyclohexane

TRICHLOROMETHANE, (CHLOROFORM, $CHCl_3$)

Chloroform is the simplest trihalogen compound in which three hydrogen atoms of methane are substituted by three chlorine atoms.

PREPARATION OF CHLOROFORM

Chloroform is prepared in the laboratory as well as on the industrial scale by heating ethyl alcohol or acetone with bleaching power or chlorine and alkali.

1. From Ethanol

The formation of chloroform form ethanol takes place as follows

(i) Oxidation of ethyl alcohol by chlorine

$$\underset{\text{ethanol}}{CH_3CH_2OH} + Cl_2 \longrightarrow \underset{\text{ethanal (acetaldehyde)}}{CH_3CHO} + 2HCl$$

(ii) Formation of chloral by the action of chlorine on acetaldehyde so formed.

$$CH_3CHO + 3Cl_2 \longrightarrow \underset{\text{chloral}}{CCl_3CHO} + 3HCl$$

(iii) Hydrolysis of chloral of chloroform by calcium hydroxide of alkali (NaOH).

$$\underset{\substack{\text{slaked lime}\\ \text{(from bleaching powder)}}}{Ca\begin{matrix} /OH \\ \backslash OH \end{matrix}} + \underset{\text{chloral}}{\begin{matrix} CCl_3CHO \\ CCl_3CHO \end{matrix}} \longrightarrow \underset{\text{chloroform}}{2CHCl_3} + \underset{\text{calcium formate}}{\begin{matrix} HCOO\backslash \\ HCOO/ \end{matrix}Ca}$$

or $$\underset{\text{chloral}}{CCl_3CHO} + NaOH \longrightarrow CHCl_3 + \underset{\text{sod, formate}}{HCOONa}$$

The net reaction when bleaching power is used is,

$$\underset{\text{ethanol}}{2CH_3CH_2OH} + 8Cl_2 + \underset{\substack{\text{from bleaching}\\ \text{powder}}}{Ca(OH)_2} \longrightarrow \underset{\text{chloroform}}{2CHCl_3}$$

$$+ \underset{\text{calcium formate}}{(HCOO)_2\,Ca} + 10HCl$$

2. From Acetone (from Propanone)

When acetone is heated with bleaching power, or with chlorine and alkali, the following reactions take place.

(i) Reaction of chlorine with acetone producing trichloroacetone.

$$\underset{\text{acetone}}{CH_3 \cdot CO \cdot CH_3} + \underset{\substack{\text{from bleaching}\\ \text{powder}}}{3Cl_2} \longrightarrow \underset{\text{trichloroacetone}}{CCl_3 \cdot CO \cdot CH_3} + 3HCl$$

(ii) Trichloroacetone is decomposed by either calcium hydroxide or an alkali to give chloroform.

$$CCl_3 \cdot CO \cdot CH_3 + NaOH \longrightarrow CHCl_3 + \underset{\text{sodium acetate}}{CH_3COONa}$$

$$2CCl_3 \cdot CO \cdot CH_3 + \underset{\text{from bleaching powder}}{Ca(OH)_2} \longrightarrow \underset{\text{chloroform}}{2CHCl_3} + \underset{\text{calcium acetate}}{(CH_3COO)_2Ca}$$

The net reaction, when acetone is heated with bleaching powder is,

$$\underset{\text{acetone}}{2CH_3 \cdot CO.CH_3} + 6Cl_2 + \underset{\text{from bleaching powder}}{Ca(OH)_2} \longrightarrow \underset{\text{chloroform}}{2CHCl_3}$$

$$+ \underset{\text{calcium acetate}}{(CH_3COO)_2Ca} + 6HCl$$

3. Preparation of Pure Chloroform

Pure chloroform needed for medicinal use is obtained by heating chloral hydrate with concentrated solution of NaOH.

$$\underset{\text{chloral hydrate}}{CCl_3CHO \cdot H_2O} + NaOH \xrightarrow{\Delta} \underset{\text{chloroform}}{CHCl_3} + \underset{\text{sod. formate}}{HCOONa} + H_2O.$$

MANUFACTURE OF CHLOROFORM

Chloroform is generally manufactured by the following methods.

(i) From methane. Chloroform can be manufactured by chlorinating methane. A mixture of $CH_4 : Cl_2 : N_2$ in the ration 8 : 3 : 80 is passed over a partially reduced sample of cupric chloride chloroform is recovered by distillation.

$$CH_4 + 3Cl_2 \longrightarrow CHCl_3 + 3HCl$$

(ii) From carbon tetrachloride. Chloroform can also be prepared on a large scale by partial reduction of carbon tetrachloride using iron filings and water.

$$\underset{\text{carbon tetrachloride}}{CCl_4} + 2H \xrightarrow{Fe/H_2O} CHCl_3 + HCl$$

PHYSICAL PROPERTIES OF CHLOROFORM

Some important physical properties of chloroform ($CHCl_3$) are.

(i) Chloroform is a heavy, colourless, sweet-smelling liquid, (b.p. = 334 K, (61°C) f.P. = 210 K, (– 63°C). Chloroform dissolves substances like oils, fats, and waxes etc. So, chloroform is a good solvent for oils, fats and waxes.

(ii) Chloroform is insoluble in water but soluble in organic liquids like ether, alcohol etc.

(iii) It is non-inflammable.

(iv) Its vapour, if inhaled causes temporary unconsciousness. Therefore, it is used as a general anaesthetic.

CHEMICAL PROPERTIES OF CHLOROFORM

Some important chemical properties of chloroform are described below.

1. **Oxidation :** It undergoes oxidation in the presence of light to give highly poisonous substance called phosgene, (carbonyl chloride).

$$2CHCl_3 + O_2 \xrightarrow[\text{air}]{\text{sunlight}} \underset{\substack{\text{phosgene} \\ \text{(carbonyl chloride)}}}{2COCl_2} + 2HCl$$

It is the presence of phosgene in long-stored chloroform which makes it unfit to be used as anaesthetic. It is for this reason that the chloroform is stored in coloured bottles filled upto the stopper

- *To prevent oxidation of chloroform :* It is very important of prevent oxidation of chloroform. This can be done as follows.

(a) Chloroform should be stored in dark-coloured bottles to cut of light.

(b) The bottle containing chloroform should be filled upto the stopper, and stoppered tightly to avoid the presence of air inside the bottle.

(c) A small quantity of ethanol (1%) should be added to chloroform at the time packaging. Ethanol converts phosgene (if, formed) to non-poisonous, and non-volatile neutral ethyl carbonate.

$$\underset{\text{phosgene}}{COCl_2} + \underset{\text{ethanol}}{2C_2H_5OH} \longrightarrow \underset{\text{ethyl carbonate}}{O{=}C(OC_2H_5)_2} + 2HCl$$

Thus, the formation and accumulation of phosgene in chloroform is prevented.

2. **Reduction :** Chloroform undergoes reduction with Zn. HCl in the presence of ethyl alcohol to give dichloromethane, (methylene chloride), but when reduced with Zn dust and water, methane is obtained.

$$CHCl_3 + 2[H] \xrightarrow[\text{ethanol}]{Zn/HCl} CH_2Cl_2 + HCl$$

$$CHCl_3 + 6[H] \xrightarrow{Zn/H_2O} CH_4 + 3HCl$$

3. **Hydrolysis :** Chloroform gets hydrolysed when boiled with aqueous solution of potassium hydroxide.

$$\underset{\text{chloroform}}{H{-}C(Cl)_3} + 3KOH \xrightarrow[-3KCl]{\text{boil}} \underset{\text{unstable}}{\left[H{-}C(OH)_3\right]} \xrightarrow{-H_2O}$$

$$\underset{\text{formic acid}}{H{-}C(=O)OH} \xrightarrow{KOH} \underset{\text{pot. formate}}{H{-}C(=O)OK} + H_2O$$

4. **With Primary Amines:** *Carbylamine Reaction* : When chloroform is warmed with a primary amine, and alcoholic potassium hydroxide, an offensive smelling compound called carbylamine (isocyanide) is formed. **Carbylamine reaction is used as a test for chloroform.**

$$\underset{\text{primary amine}}{RNH_2} + CHCl_3 + 3KOH\,(alc) \xrightarrow{\text{warm}} \underset{\substack{\text{carbylamine (isocyanide)}\\ \text{offensive smelling}}}{R \cdot N \overset{\rightarrow}{=} C}$$

$$+ 3KCl + 3H_2O$$

Example:

$$\underset{\text{aniline}}{C_6H_5NH_2} + CHCl_3 + 3KOH\,(alc) \longrightarrow \underset{\text{phenyl carbylamine}}{C_6H_5N \overset{\rightarrow}{=} C}$$

$$+ 3KCl + 3H_2O$$

$$\underset{\text{ethylamine}}{C_2H_5NH_2} + CHCl_3 + 3KOH\,(alc) \xrightarrow{\text{warm}} \underset{\text{ethyne (acetylene)}}{C_2H_5N \overset{\rightarrow}{=} C} + 3KCl + 3H_2O$$

5. With Silver Powder : On heating with silver powder at high temperature, ethyne (acetylene) is obtained.

$$\underset{\text{chloroform}}{2CHCl_3} + 6Ag \xrightarrow{\Delta} CH \equiv CH + 6AgCl$$

6. With Phenol : When chloroform is heated with phenol at 340 K, and sodium hydroxide followed by hydrolysis, salicyaldehyde, (*o*-hydroxybenzaldehyde) is formed. This reaction is called **Reimer-Tiemen reaction.**

$$\underset{\text{phenol}}{C_6H_5{-}OH} + NaOH + CHCl_3 \xrightarrow[\text{hydrolysis}]{340\text{ K}} \underset{\substack{\text{salicylaldehyde} \\ \text{(o-hydroxybenzaldehyde)}}}{C_6H_4(CHO){-}OH} + 3NaCl + H_2O$$

7. With Acetone : Chloroform condenses with acetone in the presence of potassium hydroxide to give chloretone. **Chloretone is used as a sleep-inducing *(hypnotic)* drug.**

$$\underset{\text{acetone}}{(CH_3)_2C{=}O} + CHCl_3 \xrightarrow{KOH} \underset{\text{chloretone}}{(CH_3)_2C(OH)_2}$$

USES OF CHLOROFORM

Chloroform is used,

(i) as an anaesthetic. It has now being replaced.

(ii) as a solvent for fats, waxes and resins.

(iii) as a preservative for anatomical specimen.

(iv) for the preparation of chloretone, chloropicrin etc.

TRIIODOMETHANE (IODOFORM, CHI_3)

Iodoform is another simple trihalogen derivative of methane. Iodoform is obtained by substituting three hydrogen atoms of methane by three iodine atoms.

PREPARATION OF IODOFORM

Iodoform is obtained by the action of iodine on ethanol (ethyl alcohol), of on propanone (acetone) in the presence of an alkali. This reaction is called *haloform reaction,* and is commonly known as *iodoform test.*

1. From Ethanol : When ethanol is warmed (60°C) with iodine in the presence of an alkali (NaOH, or Na_2CO_3), iodoform is obtained,

Step. 1. $NaOH + I_2 \longrightarrow \underset{\text{sod. hypoiodite}}{NaOI} + NaI + H_2O$

Step. 2. Oxidation of ethanol to ethanal (acetaldehyde).

$$\underset{\text{ethanol}}{CH_3CH_2OH} + NaOH \longrightarrow \underset{\text{ethanal}}{CH_3CHO} + NaI + H_2O$$

Step. 3. Ethanal so formed undergoes iodination.

$$\underset{\text{ethanal}}{CH_3CHO} + 3NaOI \longrightarrow \underset{\text{tri-iodoacetaldehyde}}{CI_3CHO} + 3NaOH$$

Step. 4. Hydrolysis (cleavage) of CI_3CHO by alkali.

$$CI_3CHO + NaOH \longrightarrow \underset{\substack{\text{iodoform}\\\text{(yellow ppt.)}}}{CHI_3} + HCOONa$$

Net Reaction

2. From Acetone : Iodoform can also be prepared by warming acetone with iodine in the presence of an alkali. The reactions are

Step. 1. $2NaOH + I_2 \longrightarrow NaOI + NaI + H_2O$

Step. 2. Iodination of acetone.

$$CH_3 \cdot CO.CH_3 + 3NaOI \longrightarrow \underset{\substack{\text{1,1,1-tri-iodopropane}\\\text{(triiodoacetone)}}}{CI_3 \cdot CO \cdot CH_3} + 3NaOH$$

Step. 3. Hydrolysis (cleavage) of $CI_3{\cdot}CO{\cdot}CH_3$ by alkali.

$$\underset{\text{triiodoacetone}}{CI_3 \cdot CO \cdot CH_3} + NaOH \longrightarrow \underset{\substack{\text{iodoform}\\\text{(yellow ppt.)}}}{CHI_3} + CH_3COONa$$

Net Reaction

$$\underset{\substack{\text{propanone}\\\text{(acetone)}}}{CH_3 \cdot CO \cdot CH_3} + 3I_2 + 4NaOH \xrightarrow{\text{warm}} \underset{\substack{\text{iodoform}\\\text{(yellow ppt.)}}}{CHI_3} + 3NaI$$

$$+ CH_3COONa + 3H_2O.$$

The reactions for the preparation of iodoform when Na2CO3 is used in place of NaOH are as follows.

(a) when ethanol is used.

$$\underset{\text{ethanol}}{CH_3CH_2OH} + 4I_2 + 3Na_2CO_3 \xrightarrow{\text{warm}} \underset{\text{iodoform}}{CHI_3}$$

$$HCOONa + 5NaI + 2H_2O + 3CO_2$$

(b) when propanone (acetone) is used.

$$\underset{\substack{\text{propanone}\\ \text{(acetone)}}}{CH_3 \cdot CO \cdot CH_3} + 3I_2 + 2Na_2CO_3 \xrightarrow{\text{warm}} \underset{\text{iodoform}}{CHI_3}$$

$$CH_3COONa + 3NaI + H_2O + 2CO_2$$

PHYSICAL PROPERTIES OF IODOFORM

Iodoform is a yellow crystalline solid (hexagonal plates) with characteristic unpleasant odour. It melts at 392 K, (119°C).

CHEMICAL PROPERTIES OF IODOFORM

Iodoform shows chemical behaviour similar to that of chloroform, but it is less stable. Some of the typical reactions given by iodoform are.

1. **Stability :** On heating, iodoform decomposes to give iodine vapour. Due to the liberation of free iodine iodoform shows antiseptic action. Moisture, air and light also cause the decomposition of iodoform.
2. **Hydrolysis :** On boiling with alcoholic potash (KOH (alc)), iodoform gives potassium formate.

$$\underset{\text{iodoform}}{H{-}C\begin{cases} I + KOH\,(alc.) \\ I + KOH\,(alc.) \\ I + KOH\,(alc.) \end{cases}} \xrightarrow{\Delta} 3KI + \underset{\text{(unstable)}}{\left[H{-}C \begin{matrix} OH \\ OH \\ OH \end{matrix} \right]}$$

$$\xrightarrow[-H_2O]{} \underset{\text{formic acid}}{H{-}C\begin{matrix} {/\!\!/}O \\ \diagdown OH \end{matrix}} \xrightarrow{KOH} \underset{\text{formic acid}}{H{-}C\begin{matrix} {/\!\!/}O \\ \diagdown OH \end{matrix}} + H_2C$$

3. **Heating with Silver Powder :** When heated with Ag powder, iodoform gives ethyne (acetylene).

$$CHI_3 + 6Ag + I_3CH \longrightarrow \underset{\text{ethyne (acetylene)}}{CH \equiv CH} + 6AgI$$

4. **With Primary Amines:** *Carbylamine Reaction* : Iodoform when heated with a primary amine and alcoholic potash, gives an offensive smelling compound known as carbylamine.

$$\underset{\text{aniline}}{C_6H_5NH_2} + \underset{\text{iodoform}}{I_3HC} + 3KOH\,(alc) \longrightarrow \underset{\text{phenyl carbylamine}}{C_6H_5NC} + 3KI + 3H_2O$$

5. **With Silver Nitrate:** Iodoform gives yellow coloured precipitate of AgI when heated with alcoholic silver nitrate.

USES OF IODOFORM

Iodoform is used as an antiseptic. Its antiseptic property is due to the liberated iodine. However, due to its unpleasant smell, it has been replaced by other formulations containing iodine.

IODOFORM TEST

Iodoform test is given by all the compounds which contain a methyl group attached to the carbonyl group *i.e.*, $CH_3 - \underset{|}{C} = O$ group, or the compounds that can produce $CH_3 - \underset{|}{C} = O$ group upon oxidation, *e.g.*, those containing ,

$$CH_3 - \overset{|}{\underset{\underset{OH}{|}}{C}}H = O \text{ group.}$$

Thus, *ethanol propanone (acetone), and 2-propanol (isopropyl alcohol) give iodoform test.*

Procedure : Take about 1 g of Na_2CO_3 in 5 mL of water in a boiling tube. Add about 0.5 mL of ethanol (or the given compound), and warm the solution to about 60°C on a water bath. Now, add 0.5 g of iodine gradually till a permanent yellow-brown colour is obtained. At this stage a few drops of dil., sodium hydroxide solution are added to discharge thee yellow colour. Yellow crystals of iodoform immediately separate out.

USEFULNESS OF THE IODOFORM TEST

Iodoform test can be used to distinguish between the following pairs.

Pairs of compounds	*Compound which gives positive iodoform test*
1. Ethanol and Methanol	Ethanol
2. Ethanol and Propan-1-ol	Ethanol
3. Methanol (formaldehyde) and Ethanal (acetaldehyde)	Ethanal (acetaldehyde)
4. Pentan-2one (Methyl propyl ketone), and Pentan-3-one (diethyl ketone)	pentan-2-one

Chlorine and bromine can replace iodine in iodoform to give chloroform and bromoform respectively. In this generalised form, the reaction is know as **haloform reaction.**

TETRACHLOROMETHANE (CARBON TETRACHLORINE, CCl_4)

Carbon tetrachloride is the most important tetrahalogen derivatives of alkanes.

PERPARATION OF CARBON TETRACHLORIDE

Carbon tetrachloride can be prepared on the industrial scale by the following methods.

(i) From methane: By the chlorination of methane in the presence of sunlight.

$$CH_4 + 4Cl_2 \xrightarrow{hv} CCl_4 + 4HCl$$

(ii) From carbon disulphide: Carbon tetrachloride can be obtained by the action of chlorine on carbon disulphide in the presence of aluminium chloride as a catalyst.

$$\underset{\text{carbon disulphide}}{CS_2} + 3Cl_2 \xrightarrow{AlC_3} CCl_4 + \underset{\text{sulphur monochloride}}{S_2Cl_2}$$

PHYSICAL PROPERTIES OF CARBON TETRACHLORIDE

(i) Carbon tetrachloride is a colourless heavy liquid (b.p. 350 K, (77°C)).

(ii) It is insoluble in water but soluble in organic solvents.

(iii) It is a good solvent for oils, fats and resins.

(iv) It is non-inflammable.

(v) It is stable at red heat (≈ 500°C)

In past carbon tetrachloride was used as fire extinguishing material, under the name *pyrene.* When a find spray of liquid CCl_4 is directed at the fire, it vaporises. The dense vapour of CCl_4 cut off the supply of air to the burning article. As a result, the fire gets extinguished.

However, at higher temperature in the presence of steam, carbon tetrachloride gets oxidised to phosgene. So, care should be taken while using carbon tetrachloride in fire fighting operations,

$$CCl_4 + H_2O\,(\text{steam}) \xrightarrow{500^\circ C} \underset{\substack{\text{phosgene} \\ (\text{a poisons gas})}}{COCl_2} + 2HCl$$

CHEMICAL PROPERTIES OF CARBON TETRACHLORIDE

Some chemical reactions shown by carbon tetrachloride are,

(i) *Reduction* : Carbon tetrachloride on reduction with moist iron filings gives chloroform.

$$CCl_4 + 2H \xrightarrow{Fe.H_2O} CHCl_3 + HCl$$

(ii) *Hydrolysis* : In the presence of steam, CCl_4 gives phosgene.

$$CCl_4 + \underset{(\text{steam})}{H_2O} \longrightarrow \underset{\substack{\text{phosgene} \\ (\text{a poisonous gas})}}{COCl_2} + 2HCl$$

USES OF CARBON TETRACHLORIDE

Carbon tetrachloride is used,

(i) As a solvent for fats, oils and waxes etc,

(ii) In manufacturing freon (dichlorodifluoromethane).

(iii) As a fire extinguishing agent, (with all precautions).

Carbon tetrachloride is highly toxic. Moderate exposure to it causes headache, nausea and dizziness. It harms gastrointestinal tract, central nervous system, liver and lungs. It irritates eyes and dires the skin. Prolonged exposure may cause unconsciousness or even death. It is suspected to be carcinogenic.

p-DICHLOROBENZENE

Preparation. *p*-dichlorobenzene is prepared by chlorination of benzene.

$$\text{benzene} + Cl_2 \xrightarrow[HCl]{\text{dark, iron}} \text{chlorobenzene}$$

$$\text{chlorobenzene} \xrightarrow[-HCl]{+Cl_2 \text{ (iron, in dark)}} o\text{-} + p\text{- dichlorobenzene}$$

In the mixture of o-and *p*- dichlorobenzene, the *p*- variety is present in larger amounts. The reaction proceeds through electrophilic substitution.

Properties and uses : It is a white, volatile solid (m.p. 52°C). It readily sublimes. It is used as a general insecticide, germicide, soil fumigant, deodorant and moth-repellent.

FREONS

The chlorofluoro compounds of methane and ethane are collectively called *freons*. Freons are extremely stable, nonreactive, rontoxic, noncorrosive and easily liquefiable gasses.

Dichlorodifluoromethane (CCl_2F_2, Freon 12) is one of the most common industrially used freons.

DICHLORODIFLUOROMETHANE (CF_2Cl_2)

Freon is dichlorodifluoromethane. It is a non-toxic gas which can be liquefied very easily. It is non-corrosive, and has a pleasant odour.

Preparation : It is made from carbon tetrachloride and dry hydrogen fluoride, or antimony trifluoride (SbF_3) using antimony pentachloride as a catalyst. This reaction is called *Swarts reactions*.

$$CCl_4 + 2HF \longrightarrow \underset{\text{freon 12}}{CF_2Cl_2} + 2HCl$$

$$3CCl_4 + 2SbF_3 \xrightarrow[SbCl_5]{} \underset{\text{freon 12}}{3CCl_2F_2} + 2SbCl_3$$

Uses : (i) It is used as a refrigerant in refrigerant and air conditioners.

(ii) It is used as a propellant for aerosols and foams to spray out deodorants, hair sprays, shaving foams etc.

Harmful effects : Freon is found to destroy ozone in the upper atmosphere.

The depletion of ozone layer is very harmful to the life on the earth. Efforts are being made to restrict the use of freon, and related compounds.

BENZENEHEXACHLORIDE (BHC, $C_6H_6CL_6$)

Commercially, benzenehexachloride (BHC) is known as **Lindane** or *Gammaxzne.*

Preparation : BHC is obtained by heating benzene with chlorine in presence of sunlight/ultraviolet radiation. It is a mixture of several isomers, but only the gamma isomer exhibits insecticidal/ pesticidal activity.

$$C_6H_6 \text{ (benzene)} + 3Cl_2 \xrightarrow[UV]{h\nu} C_6H_6Cl_6 \text{ (1,2,3,4,5,6-hexachlorocyclohexane)}$$

DICHLORO-DIPHENYL TRICHLOROACETALDEHYDE, (DDT)

The IUPAC name of dichloro-diphenyl trichloroacetaldehyde is 2,2-bis (p-chlorophenyl)- 1, 1, 1 - trichloroacetaldehyde.

Preparation : DDT can be prepared by heating a mixture containing chloral (trichloroacetaldehyde) and chlorobenzene (mole ratio 1 > 2) in the presence of conc. H_2SO_4.

$$C_6H_5Cl \text{ (chlorobenzene)} + CCl_3CHO \text{ (chloral)} \xrightarrow{\text{conc. } H_2SO_4} Cl{-}C_6H_4{-}\underset{CCl_3}{\underset{|}{CH}}{-}C_6H_4{-}Cl$$

Properties :(i) It is white crystalline sodium (m.p. 382 K).

(ii) It is practically odourless.

(iii) It is insoluble in water, but readily dissolves in benzene, kerosene, alcohol and freon.

(iv) It is remarkably persistent, *i.e.* DDT is non-biodegradable.

Uses : It is used as a contact insecticide. DDT is used to kill insects, *e.g.*, mosquitoes. However, its extreme chemical stability makes it potentially dangerous towards humans and animals in its long term effects.

The use of DDT has been banned in many countries.

PERFLUOROCARBONS (PFCs)

The fluorocarbons having the general formula C_nF_{2n+2} are called *perfluorocarbons.* Thus perfluorocarbons are the derivatives of alkanes in which all hydrogen atoms of the alkane molecule are replaced by fluorine atoms.

Preparation : Perfluorocarbons (PFCs) are obtained by controlled fluorination of vaporized alkanes mixed with nitrogen gas, in the presence of a catalyst.

$$C_7H_{16} + 16F_2 \xrightarrow[\text{CoF}_2\text{ (catalyst)}]{\text{vapour phase, }N_2\text{ }300°C} \underset{\text{perfluorohep tan e}}{C_7F_{16}} + 16\,HF$$

Properties : Perfluorocarbons are,

(i) colourless, odourless, non-inflammable, noncorrosive and unreactive gases, liquids or solids.

(ii) nonpolar and extremely stable, stable even to ultraviolet and other high energy radiations, and therefore, do not deplete ozone layer.

(iii) good electrical insulators

Use : Perfluorocarbons are used

(i) as lubricants, surface coatings, dielectries and heat transfer media in high voltage electrical equipments.

(ii) for vapour phase soldering, leak detection of sealed microchips etc., in electronic industry.

(iii) fin skin care cosmetics, wound healing, retina rearrachment, carbon monoxide poisoning and medical diagnosis.

Chloroform has been used in surgery as an anesthetic, and for this purpose it had to be pure. In the presence of air and light, chloroform slowly forms carbonyl chloride, which is extremely poisonous.

$$CHCl_3 + \frac{1}{2}O_2 \longrightarrow COCl_2 + HCl$$

Chlorine, water and carbon dioxide are also produced. Anesthetic chloroform was therefore kept in well-stopped dark-brown or blue bottles.

Ethanol was also added (1 per cent), but its functional is not quite clear. According to some authors, it *retards* the decomposition of the chloroform. This is supported by the fact that infra-red measurements of such mixtures show the absence of the carbonyl frequency.

A dedicate test for chloroform is the isocyanide test. This is carried out by heating chloroform with ethanolic potassium hydroxide and aniline, where by phenyl isocyanide is formed and is readily detected by its nauseating odour.

$$CHCl_3 + 3KOH + C_6H_5NH_2 \longrightarrow C_6H_5NC + 3KCl + 3H_2O$$

Chloroform is widely used in industry as a solvent for fats, waxes, resins, number, etc.

Bromoform may be prepared by methods similar to those used for chloroform, but it is prepared industrially by the electrolysis of an aqueous solution of acetone or ethanol containing sodium carbonate and potassium, bromide (acetone gives a better yield than ethanol). The solution is maintained at about 20°C, and hydrobromic acid is run in to neutralise the sodium hydroxide produced during the electrolysis .Bromine is set free at the anode, and probably reacts in the same way as does chlorine in the preparation of chloroform.

Bromoform is a liquid, b.p. 149.5°C, and smells like chloroform, which it closely resembles chemically.

Iodoform is prepared industrially by the electrolysis of an aqueous solution of ethanol or acetone containing sodium carbonate and potassium iodide (ethanol gives a better yield than acetone). The solution is maintained at 60–70°C, and a current of carbon dioxide is passed through the solution to neutralise the sodium hydroxide formed.

Iodoform crystallises in yellow hexagonal plates, m.p. 119°C. It is insoluble in water, but is readily soluble in ethanol and ether. It is used as an antiseptic, but its antiseptic properties are due to the liberation of free iodine, and not to iodoform itself. Iodoform chemically resembles chloroform and bromoform.

POLYHALOGEN DERIVATIVES

Carbon tetrachloride : CCl_4, is prepared industrially in several ways.

(i) By the action of chlorine on carbondisulphide in the presence of aluminium chloride as catalyst.

$$CS_2 + 3Cl_2 \xrightarrow{AlCl_3} CCl_4 + S_2Cl_2$$

The sulphur monochloride is removed by fractional distillation, and the carbon tetrachloride a then shaken with sodium hydroxide, and distilled.

(ii) By the chlorination of methane (see methyl chloride).

(iii) By *chlorinolysis.* This term described the process of chlorinating an organic compound under conditions which rupture the carbon-carbon bond to yield chloro-compounds with fewer carbon atoms than the original compound.

Chlorinolysis may be effected with or without a catalyst, *e.g.*, the hydrocarbon and chlorine are heated at high temperature (300–40°C) and under high pressure (about 70 kg/cm²). The product is usually a mixture, *e.g.*, propane gives both carbon tetrachloride and hexachloroethane.

$$C_3H_8 \xrightarrow{Cl_2} CCl_4 + C_2Cl_6$$

Carbon tetrachloride is a colourless liquid b.p. 77°C, which has a sickly smell. It is insoluble in water but readily soluble in ethanol and ether. Since its vapour is non-inflammable, carbon tetrachloride is widely used as an industrial solvent (for fats, oils, resins, lacquers, etc). It is also used as a fire-extinguished under the name of *Pyrene.*

Carbon tetrachloride is stable at red heat (about 5000°C), but when its vapour comes into consist with water vapour at this temperature, some carbonyl chloride is formed.

$$CCl_4 + H_2O \longrightarrow COCl_2 + 2HCl$$

Hence after using Pyrene to extinguish a fire, the room should be well ventilated.

Carbon tetrachloride is reduced by moist iron filings to chloroform (q.v). The alkaline hydrolysis of carbon tetrachloride gives the same products (formate and carbon monoxide) chloroform, but the rate of reaction is slower (Hing, 1954).

1,1,2,2-Tetrachloroethane (acetylene tetrachloride b.p. 146°C, is manufactured by the reaction between acetylene and chlorine in the presence of ferric chloride at 80°C. It is non-inflammable, and hence is widely used as solvent for oils, fats, paints, varnishes, rubber, etc.

Tetrachloroethane is the striating material for the industrial preparation of trichloroethylene (by 88°C) and tetrachloroethylene (b.p. 121°C), both are used as solvents.

$$CHCl_2CHCl_2 \xrightarrow[\text{cal.}]{300\text{—}500°C} CCl_2 = CHCl \xrightarrow[\text{light}]{Cl_2} CCl_3CHCl_2$$

$$\xrightarrow[\text{cal.}]{300°C} CCl_2 = CCl_2$$

1,2-Dichloroethylene may be prepared by the action of zinc dust on 1.1.2.2-tetrachloroethane.

$$CHCl_2CHCl_2 \xrightarrow{Zn} CHCl = CHCl$$

It is manufactured by the action of iron brings on hexachloroethane. It is a liquid, and exists as two geometrical isomers, *cis,* b.p. 60°C, *trans,* b.p. 48°C, and is used as a rubber solvent, etc.

Hexachloroethane *(perchloreoethane)*. m.p. 187°C, is manufactured by passing a mixture of s-tetrachlroethane and chlorine over aluminium chloride, and by the chlorinolysis of propane (see carbon tetrachloride, above). It is used as a moth repellent.

FLUORINE DERIVATIVES OF THE ALKANES

Most organic compounds burn or explode when treated with fluorine gas. This great reactivity a partly due to the low dissociation energy of the fluorine molecule (150-6 kJ mol^{-1}). Also, one of the termination steps is $R^{\cdot} + F^{\cdot} \rightarrow RF$. This is very strongly exothermic (447.7kJ), and the energy liberal is larger than that required to break a C—C bond (347.3 kJ), thereby resulting in fission of the organic molecule.

Aliphatic fluorine compounds may be obtained in several ways.

(i) Direct fluorination of hydrocarbons may be carried out successfully by diluting the fluorine was nitrogen and carrying out the reaction in a metal tube packed with copper gauze at a temperature 150–350°C. It is very difficult to control the fluorination, and the product is usually a complex mixture *e.g.*, methane gives CH_3F, CH_2F_2, CHF_3, CHF_3 CF_4, C_2F_6 and C_3F_8, ethane gives CF_4, C_2F_6, $CH_3CHF_4CH_2FCHF_2$, no mono- or s-difluorethane is obtained.

When catalysts other than copper (actually CuF_2) are used, *e.g.*, AgF, CoF_2, MnF_2, prefluorinationn compounds are obtained, *e.g.*, n- heptane gives perfluoroheptane.

$$2AgF + F_2 \longrightarrow 2AgF_2$$

$$C_7H_{16} \xrightarrow{AgF_2} C_7F_{16}$$

It appears that when a catalyst is used, the perfluoro-compound obtained usually has the same number of carbon atoms as the original compound, if no catalyst is used, fluoro-compounds with fewer carbon atom are usually obtained.

The mechanism of direct fluorination is still obscure, but it appears that the first step is the conversion of the catalyst into a higher fluoride, *e.g.*, AgF in to AgF_2 into CoF_3, etc. Some of these higher fluorides have been isolated, *e.g.*, AgF_2.

(ii) Alkenes and alkynes add on hydrogen fluoride under pressure to form fluoro under pressure to form fluoro-derivatives of the alkanes, *e.g.*,

$$C_2H_4 + HF \longrightarrow C_2H_5F \text{ (f.g. g)}$$

$$CH_3C \equiv CH + 2HF \longrightarrow CH_3CH_2CH_3 \text{ (f.g.g)}$$

If the unsaturated compound contains a halogen atom (other than fluorine), this atom may be replaced by fluorine, *e.g.*,

$$RCCl{=}{=}CH_2 + HF \longrightarrow RCFClCH_3$$

$$RCCl{=}{=}CH_2 + 2HF \longrightarrow RCF_2CH_3 + HCl$$

Lead tetrafluoride (from lead dioxide and hydrogen fluoride) is particularly useful for the addition of two fluorine atoms to an alkene containing chlorine, *e.g.*,

$$CCCl_2 = CCl_2 + PbF_4 \xrightarrow{25°C} CFCl_2CFCl_2 + PbF_2.$$

It is also possible to add fluorine directly without a catalyst to highly halogenated alkenes ,*e.g.*,

$$CFCl{=}CFCl + F_2 \longrightarrow CF_2ClCF_2Cl$$

(iii) Fluorine compounds may be prepared indirectly by heating organic halides with inorganic fluorides such as AsF_3, AgF, $Hg2F_2$, etc, *e.g.*,

$$C_2H_5Cl + AgF \longrightarrow C_2H_5F + AgCl \text{ (ex)}$$

This method was first used by Swarts (1898), and so is known as the Swarts reactions.

When the organic halide contains two or three halogen atoms attached to the same carbon atom, the best yield of fluoride is obtained when CoF3 is used, but SbF3 gives yield almost as good (and is more accessible). *e.g.*,

$$3CH_2CCl_2 + 2SbF_3 \longrightarrow 3CH_3CF_2CH_3 + 2SbCl_3 \text{ (v.g..)}$$

Alternatively, hydrogen fluoride may be used under pressure in the presence of catalyst, *e.g.*,

$$CCl_4 \xrightarrow[(C.\,FeCl_3)]{HF,300°C} CF_2Cl_2 + CFCl_3$$

(iv) Another method of fluorination is the direct electrolytic method. Nickel electrodes are used, and electrochemical fluorination takes place at the anode, the reaction being carried out by the electrolysis of a solution of the organic compound in anhydrous hydrogen fluoride, *e.g.*,

$$CH_3CO_2H \longrightarrow CF_3COF$$

$$C_2H_5OC_2H_5 \longrightarrow C_2H_5OC_2F_5$$

The particular merit of this method is that is usually leaves untouched many types of functional groups. Sulphur tetrafluoride (Smith, 1962) and phenylsulphur trifluoride, $PhSF_3$ (Sheppard, 1962) are capable of replacing *oxygen* atoms by fluorine, *e.g.*,

$$RCO_2H \xrightarrow{SF_4} RCF_3\,;\; R_2CO \longrightarrow R_2CF_2;$$

$$ROH \longrightarrow RF$$

Bromine trifluoride is also useful for replacing oxygen atoms (Stevens, 1961).

The lower n-alkyl fluorides are gases. The first four members are stable, and the higher members tend to decompose spontaneously into alkene and hydrogen fluoride, *e.g.*,

$$CH_3CH_2CH_2CH_2CH_2F \longrightarrow CH_3CH_2CH_2CH{=\!=}CH_2 + HF$$

Secondary and tertiary alkyl fluorides are so unstable that it is impossible to prepare them free from alkene.

1,2-Dichlorides are also usually unstable, *e.g.*, ethylene difluoride decomposes spontaneouly 0°C into hydrogen fluoride and butadiene.

$$2CH_2FCH_2F \longrightarrow 4HF + CH_2{=}CH{-}CH{=}CH_2$$

Alkyl fluorides are readily hydrolysed by strong acids to the corresponding alcohols, alkalis not effect. On the other hand, ethylene difluoride is immediately hydrolysed by water to glycol

$$CH_2FCH_2F + 2H_2O \longrightarrow HOCH_2CH_2OH + 2HF$$

Fluorides with or three fluorine atoms on the same carbon atom are stable to water and acids, *e.g.*, CHF_3. CHF_2CHF_2.etc.

Alkyl fluorides do not react with sodium, *i.e.*, do not underego the Wurtz reaction, and do not for Grignard reagents. An intersecting compound is trifluomethyl iodide, CH_3I. It is converted fluoroform. CHF_3, by potassium hydroxide, and it combines directly with many non-metals such As, Sb. S. Se to give,, *e.g.*, with phosphorus, $(CF_3)_3P$. $(CF_3)_2PI$ and CF_3PI_2.

Chlorofluoro-derivatives of methane and ethane are used as refrigerants and for air conditions under the name of *Freons.* Which are prepared by the action of hydrogen fluoride on carbon in chloride, chloroform and hexachloroethane.

Tetrafluoroethylene, b.p. – 76, is prepared by the action of antimony trifluoride and hydrogen fluoride on chloroform, and then heating the chlorodifluoroformethane so produced at 800°C.

$$CHCl_3 \xrightarrow[HF]{SbF_3} CHF_2Cl \xrightarrow{\text{Sod C}} C_2F_4 + 2HCl \text{—other products.}$$

When tetrafluoroethylene is polymerised, the plastic. Teflon is produced. Teflon is difficult to work is inert to chemical reagents, even to boiling aqua regia.

Polychlorofluoroethylenes are valuable as oils and greases. Perfluoroheptane is useful in product for the separation of uranium isotopes by gaseous diffusion.

The C—F bond is much stronger than nay other C—X bond and increases in strength as the number of fluorine atoms attached to the carbon atom increases (CH_3F. CF_4, 485 kJ). It is largely because of this that *perfluoro* compounds are stable to heat and very reason to attack by chemical reagents.

SOLVED EXAMPLES

Example. 1:

Give the names of the major products obtained when bromoethane is treated with,

(a) an aqueous alkali

(b) sodium metal in dry ether

(c) an alcoholic potassium hydroxide solution

(d) magnesium followed by ethanol

(e) aminoethane

Solution:

The reactions are summerized below.

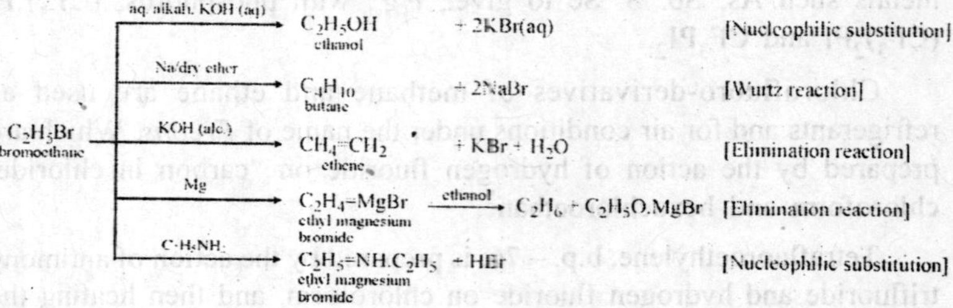

Example. 2:

With the help of chemical equations, show how you will convert 2-bromopropane to 1-bromopropane in two steps.

Solution:

$$\underset{\text{2-bromopropane}}{CH_3-\underset{\underset{Br}{|}}{CH}-CH_3} \xrightarrow[-KBr]{KOH\ (alc.)} \underset{\text{propene}}{CH_3-CH=CH_2} \xrightarrow[\text{peroxide}]{+HBr} \underset{\text{1-bromopropane}}{CH_3-CH_2-CH_2Br}$$

Example 3:

What happens when 2-bromopropane is treated with

(i) H_1 in the presence of red pphosphorus

(ii) H_2 gas in the presence of nickel as catalyst

(ii) silver nitrite

(iv) ammonia

Solution:

The reactions are summarized below.

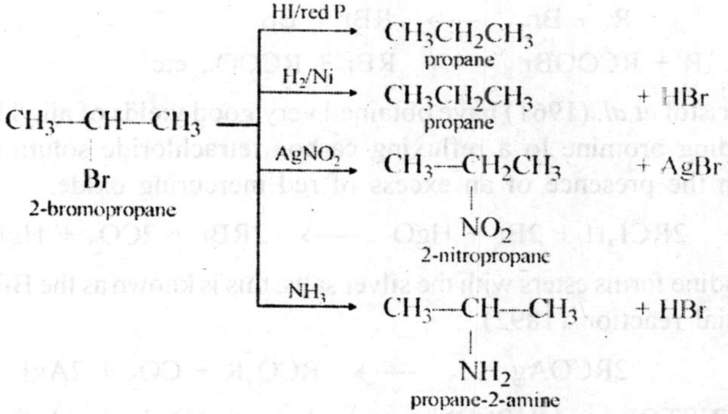

$$CH_3\text{—}\underset{\displaystyle Br}{\underset{|}{CH}}\text{—}CH_3 \;\text{(2-bromopropane)}\begin{cases}\xrightarrow{HI/\text{red P}} CH_3CH_2CH_3 \;\text{(propane)}\\ \xrightarrow{H_2/Ni} CH_3CH_2CH_3 \;\text{(propane)} + HBr\\ \xrightarrow{AgNO_2} CH_3\text{—}\underset{\displaystyle NO_2}{\underset{|}{CH}}CH_3 \;\text{(2-nitropropane)} + AgBr\\ \xrightarrow{NH_3} CH_3\text{—}\underset{\displaystyle NH_2}{\underset{|}{CH}}\text{—}CH_3 \;\text{(propane-2-amine (a primary amine))} + HBr\end{cases}$$

Example 4:

With the help of chemical equation, show how you will convert propan-1-ol to 2-bromopropane in two steps.

Solution:

$$\underset{\text{propane-1-ol}}{CH_3\text{—}CH_2\text{—}CH_2OH} \xrightarrow{\text{dhydration}} \underset{\text{propene}}{CH_3\text{—}CH=CH_2} \xrightarrow{HBr} \underset{\text{2-bromopropane}}{CH_3\text{—}\underset{\displaystyle Br}{\underset{|}{CH}}\text{—}CH_3}$$

Example 5:

With the help of chemical equations, show how you will convert 1-bromopropane to 2-bromopropane in two steps.

Solution:

$$\underset{\text{propan-1-ol}}{CH_3\text{—}CH_2\text{—}CH_2Br} \xrightarrow[\text{-KBr}]{\text{KOH (alc.)}} \underset{\text{propene}}{CH_3\text{—}CH=CH_2} \xrightarrow{+HBr} \underset{\text{2-bromopropane}}{CH_3\text{—}\underset{\displaystyle Br}{\underset{|}{CH_2}}\text{—}CH_3}$$

$$RCO_2Ag + Br_2 \longrightarrow RBr + CO_2 + AgBr$$

The yield of halides is primary > secondary > tertiary, and bromine is generally used, chlorine giving a poorer yield of alkyl chloride. The mechanism is uncertain, but a favoured theory that the first step is the formation of an acyl hypohalite which then decomposes into free radicals. The details of the later steps are less certain.

$$RCO_2Ag + Br_2 \longrightarrow RCOOBr + AgBr$$

$$RCOOBr \longrightarrow Br. + RCOO. \longrightarrow R. + CO_2$$

$$R. + Br_2 \longrightarrow RBr + Br.$$

$$R + RCOOBr \longrightarrow RBr + RCOO., \text{ etc}$$

Cristol *et al.* (1961) have obtained very good yields of alkyl bromide by adding bromine to a refluxing carbon tetrachloride solution of the acid in the presence of an excess of red mercuring oxide.

$$2RCI_2H + 2Br + HgO \longrightarrow 2RBr + 2CO_2 + HgBr_2$$

Iodine forms esters with the silver salts, this is known as the Birnbaum-Simonial reaction (1892),

$$2RCOAg + I_2 \longrightarrow RCO_2R + CO_2 + 2AgI$$

CRISTOL *et al.* (1964), however, have shown that acids form alkyl iodides if iodine and red mercûric oxide are used.

On the other hand, Bachman *et al.* (1963) have shown that aliphatic carboxylic acids may be converted into esters and alkyl halides as follows.

$$RCO_2H \longrightarrow (RCO_2)_4Pb \xrightarrow{X_2} RCO_2R + 2RX$$

Alkyl chlorides may be prepared without simultaneous ester formation by the action of lead tetracetate and lithium chloride on the carboxylic acid in benzene solution (Kochi, 1965). This reaction applicable to primary, secondary and tertiary acids, and occurs without rearrangement, *e.g.*,

$$Me_3CCH_2CO_2H \longrightarrow Me_3CCH_2Cl$$

The reaction is believed to proceed by a free-radical mechanism.

Alkyl halides may be prepared by methods involving the use of phosphorus compounds.

EXERCISES

1. Write out the structures and IUPAC names of all the dichloro-derivatives of (i) butane. (ii) isobutane.
2. Distinguish between ethylene dichloride and ethylidene dichloride.

   (i) chemically. (ii) by one physical method.
3. Draw all possible stereoisomers for.

   (i) $MeCHBrCH=CH_2$.

   (ii) $MeCHOHCH=CHMe$.

(iii) $MeCHCCO_2H$,

(iv) $BrCH_2CH_2CO$.

(v) $Me_2C(OH)CO_2H$.

4. Explain the following reactions.

(i) n-BuBr + KCN $\xrightarrow{EtOC-H_2O}$ n-BuCN

(ii) t-BuBr + KCN $\xrightarrow{EtOC-H_2O}$ $Me_2C{=}CH_2$

5. Explain the following reactions.

(i) EtOH + HI $\longrightarrow$ EtI (ii) EtOH + HCN $\longrightarrow$ no reaction

6. Calculate ΔH^θ for the reactions.

$$C_2H_6 + Br_2 \longrightarrow C_2H_5Br + HBr$$

Is the reaction thermodynamically favcurable? If so, will it proceed in the dark at room temperature. If as why not?

7. Treatment of $Me_3CCH{=}CH_2$ and $Me_3CCHOHMe$ with concentrated hydrochloric acid gives the term two isomeric alkyl chlorides. What are these two products? Explain.

8. Complete the following equations.

(i) $ClCH_2CH_2CH_2Br + KCN \xrightarrow[heat]{EtOH-H_2O}$?

(ii) PbCHO $\xrightarrow{SF_4}$?

(iii) $BrCH_2CH = CHCO_2Me \xrightarrow{AgF}$?

(iv) isoPrBr $\longrightarrow$ PrH

(v) $Me_2C = CH_2 + HCl \xrightarrow{petroxide}$

8. $EtCH_2CHBrMe \xrightarrow[(-HBr)]{EtOK} \underset{A}{1-ene} + \underset{B}{2-ene} + \underset{C}{2-ene}$

What are A, B and C and which one predominates? (Assume the mechanism is E2).

10. Discuss the reaction : $ClCH_2CHCl_2 \xrightarrow{OH^-} CH_2 = CHI_2$

11. The hydrolysis of n-BuCl in aqueous ethanol is accelerated in the presence of NaI. Explain and draw the energy profiles for the uncatalysed and catalysed reactions.

12. What is the simplest alkane that is optically active?
13. The NMR spectrum of the compound $C_3H_5Cl_2$ showed two signals, one a doublet and the other a quadrat. What is the structure of the compound?
14. State the number of signals and their multiplicity for the NMR spectra of.

 (i) CH_2F_2. (ii) Me_3CCH_2Cl.

 (iii) Me_3CCl. (iv) $ClCH_2CHClCCl_3$.

 (v) $ClCH_2CCl_2CH_3$.
15. The table gives some of the peaks in the mass spectra of two isomeric s straight chain alkyl halides. A and B. One is a primary halide. What is this one? Give your reasons.

m/e	41	43	57	85	135	164	166
(A%)	38	100	22	49	50	2.3	2.3
B(%)	38	100	12	57	1	1	1